MO41

The Bombshell Before Roswell

The Case for a Missouri 1941 UFO Crash

Paul Blake Smith

W & B Publishers

USA

MO41 The Bombshell Before Roswell © 2015.
All rights reserved by Paul Blake Smith.

W & B Publishers

For information:

W & B Publishers

9001 Ridge Hill Street

Kernersville, NC 27284

www.a-argusbooks.com

ISBN: 9781942981534

Book Cover designed by Dubya

Printed in the United States of America

"They said it was an "outer space ship" that crashed, with "little people" on board. They didn't use the terms "UFO" or "aliens" back then. But I remember people talking about it at the time, when I was ten years old in 1941," the elderly woman from just outside of Cape Girardeau told this author in 2014, *"and I was told it crashed on a farm not too far away."*

"...1939, two years before a captured UFO," wrote Dr. Edward Teller in a later-leaked letter to President Ronald Reagan in the 1980s.

"My sister's friend had been in contact with local nurses from the hospital," an elderly man from Cape Girardeau e-mailed this author in 2014, *"and he told her they turned white with fright because they had never seen anything like it before,"* when viewing the dead aliens recovered from the crash-landed spaceship.

"There's more to this story than you think," broadcaster Rush Limbaugh, a Cape Girardeau native, allegedly told a caller to his national radio program on the subject in 2005.

MO41, The Bombshell Before Roswell:
The Case for a Missouri 1941 UFO Crash

Table of Contents

Preface

"Is mankind alone in the cosmos? Is Earth being visited by intelligent creatures?" Those are the epic, controversial questions many human beings have asked and pondered for millennia. Perhaps, just *perhaps*, there really *have* been genuine sightings of alien beings by this world's citizens at some point in the past, and perhaps also there have been vehicular accidents here on planet Earth by those extraterrestrials, within man's millions of years of existence. Who knows for sure? But what we *do* know is that inside the boundaries of the United States of America, from its very inception in 1776, freedom of speech and the power of the press have been very substantial and prevalent. Frankly, so has gossip and generational storytelling. Word of mouth and worth of media, they have been part of the bedrock of American society since the beginning. And in its four centuries of existence, no one in America has recorded or put forward any sort of serious evidence; detailed, believable first-hand descriptions; unshakeable images; or corroborating claims for genuine sightings of visiting alien beings *until the 1940s*.

In 1897, a particular, peculiar allegation was rumored and written about to a small degree, regarding a possible alien crash-landing accident near rural Aurora, Texas. Those who have investigated this topic with today's modern forensic methods have found it very much lacking in proof and validity, with most concluding it to be mere "tall Texas tales" created within shabby "yellow journalism" designed to sell a few extra newspapers. In fact, two respected, well-known contemporary UFO investigators - very eager for genuine otherworldly mysteries to trumpet profitably to the stars in today's media - have stated flat out to their great disappointment: *the 1897 Texas affair was not genuinely extraterrestrial in nature.* It was an explainable metal-clad blimp or balloon, if it was anything physical or material at all.

So... that leaves us with the next in line: *1941 Missouri.* America's very first UFO crash controversy, the granddaddy of them all, apparently. Something truly worth researching and cataloguing in book form, for sure.

Until now, the allegation of a "non-terrestrial" crash near Cape Girardeau, Missouri, hasn't been fully explored in person and in print by anyone, anywhere. Incredibly, no one has greatly researched, written, and published a comprehensive, tangible, readable book on this potentially explosive, world-affecting subject - until now. If the fascinating claims about what went down (literally) in a farm field outside of Cape Girardeau during the pre-WWII Franklin Roosevelt administration can ever be indisputably true someday, this tale becomes of course one for the history books. As of right now, before you read this book, it seems to be the greatest story never told. At least not fully. Let this publication in your hands formally and officially overturn that notable oversight at last. It is focused solely on what first took place in this writer's upstanding hometown of Cape Girardeau and its humble surrounding countryside, and also the federal government's reaction to it all in the murky aftermath. I have researched the subject matter of this book for years, polished it repeatedly with updated information and sources as I went along, and included only small dollops of reasonable conjecture and opinion along the way. Since the story greatly revolves around a remarkable 1941 claim made by an unremarkable Cape Girardeau citizen, I have personally

tracked down this man's home and toured it, along with
what passes today as his old workplace, including its
private, locked records room. I have even traveled to this
person's final resting place, a cemetery across the state, and
received more details on his life from a surviving relative.

I have read many a related book, online website, blog,
magazine, phone directory, library publication, and
newspaper archive, searching for clues. I've placed some
phone calls, written polite letters, typed Facebook
messages, scoured computer resources, interviewed people,
and traveled about the countryside, searching around and
around, patiently seeking data, and an evident "ground
zero" for the story. I've been allowed into the inner
sanctum of Cape's esteemed local newspaper office to
inspect their digital files and old photographs. I've been
given special access to the town's old fire and police
department headquarters, now a nice city museum. I've
been allowed into a key eyewitness's home, uniquely. And
I've tried to contact and communicate with many a so-
called "UFO expert" or author on the topic, with some
politely replying and others simply ignoring me. A few of
my own fellow Cape Girardeans - and even my own
relatives! - responded much the same way. Some are
fascinated, but not necessarily believing. Others want
nothing to do with the entire topic. It irritates and/or scares
them, for various reasons. Some talk freely, others not at
all. Perhaps that is an accurate representative slice of the
global pie. From what I have experienced, opinion on the
ET crash theory in southeast Missouri seems rather divided
in half. First, about half the citizens in my old hometown I
spoke to had no idea the crash was alleged to have
occurred. Second, of those that *had* heard or read about the
claim, perhaps about half believe it, and half don't. And
roughly half say they want to hear or read more about it in
detail, while the other half don't want another word wasted
about such "nonsense."

The account herein has proven very difficult to explore and learn more. Nearly all primary sources are long since deceased, and even secondary sources through the following generation have been dying off. Old official files, letters, and records grew decrepit and yellow with age; they were long ago tossed out by various departments and offices when the digital age of computers took over people's personal and professional lives. Hard physical proof may well still exist, yet never seems to quite surface. And to make matters worse, the aging incident and its under-documented aftermath were very much hushed up at the time, with many of those in the know in 1941 largely afraid to talk, understandably, even to trusted family and friends, even *decades* after the fact.

Sources and narratives, trivia and opinion are all included in the back of this book. It is hopeful that readers will be able to absorb it all and then hopefully some will be able to fill in a few blanks with their own angles on the various stories herein by contacting the author via the publisher. Together, we can produce still more data and proof that the most stunning allegation of all time is valid and worthy of greater world respect and renown. There is enough information already being amassed as I write this that will fill a *Volume Two* someday. As for now, I thank you in advance for giving this edition your time and consideration - and hopefully your belief.

Paul Blake Smith
Spring 2016

Introduction

"Not another book about Roswell!" That might understandably be your first reaction here, but relax. Please be assured that this publication mentions only in passing at times the controversial 1947 Roswell, New Mexico, aerial crash allegations. What this unique book *is* all about is something similar, however, a subject that has previously strangely evaded a full-length hard copy book to examine all the details of what could well be one of the biggest, most amazing stories in human history. To fully understand the staggering southeastern Missouri matter at hand, however, we first need to go back and cover some Roswell ground briefly in this opener to properly set the stage.

According to many Roswell crash eyewitnesses - too numerous to detail here - in early July of 1947 certain members of the United States Army came roaring into New Mexico and swiftly scooped up something quite unusual and classified top secret. It was said to be the remains of dead gray aliens near their crash-landed spaceship (possibly *two*), and all the remaining metallic debris laying on the dusty soil. *These were not soldiers from the nearby Roswell Army airbase.* They were *outsiders*, American military men who also worked to round up and browbeat people who saw, heard, or knew too much in various locations around the so-called "Land of Enchantment." Evidently they were hard-nosed U.S. Army Counter Intelligence Corps officers. They quickly and slickly squelched the shocking story around Roswell, in the local media, and in the national press as well. They had wooden boxes ready on the ground near the crash site very swiftly, for covertly crating recovered artifacts. They had trucks and soldiers in efficient teams fanning out across the desert, for rounding up loose ends. And they employed these outside officers in area villages and even at the Roswell base very speedily, gathering more facts and shaping the situation and the official story to their liking. They were all there so rapidly, in fact, they might well have been in this general area *before* the infamous Roswell crash evidence was reported to the local law enforcement and military base.

The late Colonel Phillip Corso, in his flawed but fascinating bestseller and History Channel special, "*The Day After Roswell*," stated the U.S. Army had previously formed an elite recovery Counter Intelligence Corps unit that was very well trained in how to properly handle these sticky crash situations. In the mid-summer of 1947 they did their jobs reasonably well. What to do, who to see, and where to go, ready at a moment's notice. How to find, pick up, box, and ship out the physical evidence, and who should gain access to examining it later. How to create a cover story, through a variety of means and methods. And how to get news reporters, law officers, government workers, innocent bystanders, and even other loyal United States Army personnel to patriotically cooperate in shushing the real story "for reasons of national security."

Phil Corso and still other sources described this fast-moving, well-educated, ruthless crew that was first tipped off to some sort of unnatural aerial presence on July 1st and 2nd, 1947, by utilizing various U.S. airport and military airfield radar returns. These showed speedy foreign radar blips skipping across the country, supposedly by an incoming damaged alien spaceship, hurtling through our national airspace before it smashed into the dry New

Mexico desert floor. That's how this specially-prepared Army CIC cleanup team knew in advance where to go so quickly for the top secret, highly-classified "black op" of covert recovery and conspiratorial cover-up.

So the resulting question from all of this blares for us today: *how and why did this high-level unit get put together in the first place?* In other words, what previous "past experience" did they have to learn by and train by for future events? What served as a template, and caused the United States Army to select and uniquely educate, train, and deploy these trusted personnel? How could they be so very ready to shamelessly whitewash future aerial crashes and deny the American people the amazing truth? And *why?* The answer for these very important questions lies within this book. In a nutshell: *Cape Girardeau, Missouri. Mid-April of 1941. Before* national radar installations, the CIA and NSA (and NASA). Before even America's entrance in the ongoing Second World War.

Back in '41, the "Show-Me State" UFO crash story was seemingly going to be the most amazing nonfiction saga of all time, or at least until that point. A real-life stunner in America's Heartland that might have touched off some amount of initial mild panic in people who were unprepared for the shocking truth about advanced alien life forms, visiting here mostly undetected. "MO41" it has been duly dubbed here in shorthand for obvious reasons, a seemingly blockbuster event that *should* have penetrated the hearts and minds and imaginations of every American who heard it, and perhaps every other citizen of our troubled, war-torn planet back then, and to this day. MO41 would have literally changed the world had it made the papers and radio in its day, but. the dramatic global banner headlines and detailed accounts never made it off the ground. It was all so effectively covered up and classified at the time it was barely whispered about by a few

Americans and nearly forgotten by all, and incredibly, hasn't been told in an investigatory full-length book form. Until now.

{Before continuing, it should be noted that our subject matter is what so many commonly refer to as a "UFO" incident. Thus that specific phrase is utilized herein, even though we all know that "Unidentified Flying Object" is technically unfit and inaccurate. What smashed into mid-American terra firma was clearly identified as an *extraterrestrial spaceship*. An *alien aircraft*. It really wasn't "unknown" or "unidentifiable." Since most folks just say "UFO" as acceptable shorthand for all alien encounters, it is grudgingly used throughout this book.}

Within the pages of this book, we'll take a very close look at the first and overall best description of the alien accident, and the resulting haunting photograph it apparently produced. We'll dig into a modern Kindle e-book's allegations of one family (from Sikeston, MO) and their unique insight into the 1941 crash, and also pore over Internet clues that definitely open some eyes as well. We'll travel to Ohio to learn a startling ET crash saga connection that reaches into the depths of the U.S. Capitol Building, of all places. We'll go over some authentic, leaked, once-classified government documents that reference the stunning ET topic, and cover all the coverage: the many minor literary investigations into the Missouri incident and its tentacles will be examined.

The actual MO41 alien crash-landing was only initially rumored by some around a small Midwestern town or two, then resurfaced in a simple letter to a rather obscure UFO researcher in the 1990s. He passed it along to a better-known investigator. *He* printed the story within the chapter of a book that didn't gain mainstream publicity or "go viral" as folks say nowadays. A few more chapters or mentions within a few more UFO books brought a bit more attention as the story slowly spread. It picked up steam on the Internet and in some television news reports, more so after a cable TV documentary ran it just after the turn of the century, yet "the bombshell before Roswell" stubbornly remains murky and mysterious, under-investigated and unappreciated.

Obviously, President Franklin D. Roosevelt in the spring of 1941 could not speak openly about the staggering Cape Girardeau finds. It was hushed from the start, likely under his own orders. But he was of course worried about the possible consequences of the otherworldly evidence. No earthly nation had threatened or attacked America, but with Nazi warfare going in Europe, the increasingly ill American president had to do *something* to secure our country's airspace. The military couldn't watch the land and the skies in every county within every state... but *civilians* sure could. They could keep their eyes on foreign airships, crashes on the ground, and possible acts of aggression by foreign powers, be they aggressor nations or inquisitive cosmic visitors. Average citizens could then report any UFO sightings or accidents to the Army, as they did near Cape Girardeau. This might well be why on May 20th, 1941 - over six months before Pearl Harbor and our nation's entry into WWII - FDR suddenly issued an official proclamation creating the Office of Civil Defense and the Civil Air Patrol in the weeks after MO41 (see Chapter Six). And why he took to the national airwaves a week later to deliver a dramatic major radio address mentioning this act, heard by a whopping seventy percent of U.S. homes, the second highest of his presidency. "I have tonight issued a proclamation that an *unlimited national emergency* exists and requires strengthening of our defense to the *extreme limit* of our national power and authority," the Commander-in-Chief declared. There was "a war for world domination" going on, FDR reminded all listeners, and to ignore it would "place the nation at peril." Roosevelt spoke of civilian watchfulness and responsibility, but mainly this theme: "We are placing our military forces into strategic position. We will not hesitate to use our armed forces to repel attack." Such a general warning could certainly apply to anyone listening; people from this world, and people from beyond it. And yes, much of Roosevelt's speech was about *manmade* war, and the U.S. need for defensive measures and preparedness against "Hitlerism," in case of an assault of some kind. But we now realize something else entirely was possibly worrying FDR and his military advisors. Possible alien probes or scouts and unapproved, unannounced visitation. and even an imagined non-terrestrial *invasion*?

Historians recount how Roosevelt wanted to give this critical peacetime speech weeks earlier, but he underwent intestinal problems and then a grave anemia situation that required actual secret blood transfusions. This on top of all of his other usual physical problems, poor health wrecked much of FDR's month of May, 1941. In fact, the president didn't even leave his bed to go to the Oval Office during a critical *two weeks* of that problematic month. But he had time to think about the overseas war reports handed to him, and the findings of initial covert inquiries into the MO41 recoveries and their possible meaning, and then plotted how to take responsible protective action without setting off a panic. Thus, we have the cover art for this astonishing book for this astonishing period in American history, one that we can more fully understand today. The mysterious alien body photo of unknown origin – is it fake, or fact? - that greatly resembles what an eyewitness said she saw later in a MO41 snapshot (see Chapter Two), presented alongside a grim President Roosevelt, giving his nationwide late-night radio warning before delegates of almost every nation from North to South America, hinting as much as he dared at being ready for what might possibly come next from above.

This book seeks to explore more on President Roosevelt, and to tell the stunning MO41 affair in detail as calmly and open-mindedly as possible, to make a case that lets the reader act as the judge of its merits. Admittedly, the physical, tangible evidence to prove the case is still frustratingly nonexistent today, but many a court case is won on gathered circumstantial evidence alone. I feel I can reasonably do so herein. I can present some witness testimony; various strong clues; and some fascinating second hand hearsay but supportive evidence. I can also submit some balanced conjecture, some telling documents, and some cold hard facts on topics that may be interrelated. *But we also have to remember to open our minds and hearts to fully absorb the entire drama before coming to a conclusion and rendering a verdict.*

The full MO41 case is now ready, its presentation to the jury is as follows, and the court of public opinion is now in session. Thus, we begin.

CHAPTER ONE

What the Preacher Saw

"I'll tell you this once, and then I'm never going to speak of it again."

The Reverend William Guy Huffman, Sr., age 52, was a mild-mannered associate pastor at a struggling Baptist church in the sleepy southeastern Missouri town of Cape Girardeau, way back in the spring of 1941. He remains the original and best overall source of one truly amazing story, to say the least. Reportedly a kind and trusted man who potentially had everything to lose and nothing to gain by his extraordinary claim, Minister Huffman was unexceptional in his appearance, a bit squat and balding in his middle age, and quite devoted to his faith and his family. He didn't have much else. William was reportedly run-of-the-mill in his demeanor, his home life, and his professional duties. "Quiet" was how one relative remembered him, in a word. Even-tempered, friendly, and "hardly prone to fantasy," were other terms that come to mind. Fire and brimstone sermons to rattle a congregation were definitely not William's style. Ironically, low-key W. G. Huffman produced perhaps the most stunning and gripping discourse any preacher in American history has ever seriously uttered, perhaps any U.S. citizen at all. It was originally intended for just three people to hear, yet could go down in human history as perhaps the most sensational saga of all time - if its implications are ever proven true.

"It took a lot to shake my grandfather," his granddaughter Charlette Huffman Mann recalled around the year 2000 of patriarch William, a man at peace with his humble lot in life. But had he suddenly gone mad one night in mid-April of 1941? Was he drugged, or drunk? Or quite sober but simply honestly mistaken? Or experiencing a mental health crisis? A deeply affected Pastor Huffman somberly informed his family six years before the infamous "Roswell affair" that he personally inspected on a nearby farm a rounded, crashed spaceship and three dead alien occupants, little gray entities visiting Earth from another world. Proof positive at last that mankind is not alone in the cosmos. And he really meant it. Today, looking back over more than seven decades, it appears more and more likely that *Reverend Huffman was telling the unvarnished truth.*

William G. Huffman was born in 1888 and raised in Lebanon, Missouri, across the state from Cape Girardeau. As a young man he went to a religious school and a seminary not far from Lebanon. He became an ordained Southern Baptist minister, as well as a mild-mannered schoolteacher at times over the coming decades. William met and courted Miss Floy M. Peters in Lebanon. They ceremoniously vowed to stick together in good times and bad, then as a happily married couple moved about the Midwest in bringing God's word - and hopefully contributions - to various communities. Pastor Huffman was brought into Cape's Baptist society from Arkansas in late 1940 initially for one main purpose: to help raise much-needed funds and attendance for his adoptive Red Star Tabernacle, his specialty. William's redheaded wife and unmarried young adult son went with him, trusted him, and loved him. They also learned to help him politely raise money in small amounts. Early 1941 articles in *The Southeast Missourian* newspaper - the sturdy Cape Girardeau daily - indicated that clergyman Huffman was in the community but would not officially sign up for pastoral duties until that July at the Red Star headquarters, located on downtown Cape's once-busy North Main Street. During his first months in town, William had a critical part-time job at the tabernacle: stem the tide of red ink, much of which was caused by the tide of brown, brackish floodwaters from the nearby chocolate-brown Mississippi River. It seems the hulking Red Star Tabernacle was made of wood, susceptible to the very humid river valley air to begin with, and starting to show rot in its foundation from the frequent smelly backwater that seeped into the local sewer systems and vulnerable basements. The vulnerable structure wasn't really all that old but sat at the bottom of a hill, to make nettlesome flood problems even worse. Every few years, it seemed, the river waters would crest at such a high level - in the days before a protective cement flood wall was implemented downtown - that it would seep in and pour directly into the church's lowest levels. Thus raising building repair funds and enthusiasm was of prime importance to weary Red Star leaders. Hence the call for experienced William Huffman's charm and experience. The scenic old town of Cape Girardeau (1941 population of 20,000) perched so interestingly close to the famous, dangerous river. perhaps too close for many in what locals call "the Red Star District." Today the city is best known as the hometown of radio broadcaster Rush Limbaugh, and Major League Baseball catcher A. J. Ellis, and some of rock star Sheryl Crow's relatives. Back in 1941, it was mostly known for making shoes in the large factory down the street and train tracks from Reverend Huffman's church. And perhaps for other factories, churches, rose-filled parks, and the ever-present river.

Perhaps in some ways mild-mannered Pastor Huffman wasn't the most popular man in Cape. His job was to gently put the bite on citizens, to creatively come up with badly-needed money to cover various church repairs and projects. It was up to kindly William to not only appeal to the congregation, but to others in town, seeking out citizens with influence and money to bring in more badly-needed revenue. During this time in Cape, William and Floy and their son Wayne lived in a nice but small Red Star-rented home up the street from the church. It was literally high and dry, situated atop a hill, away from the dangerous undercurrents of the often fast-flowing Mighty Mississip, even during spring flooding season.

Finances, reconstruction work, attendance, the collection plate. inspiring others to give generously. that was what generally occupied William's mind in those springtime days. As in other jobs, the kindly Heartland clergyman raised contributions and spirits for only a few years in Cape Girardeau (1941-1944) before moving on to ply his trade in Kingfisher, Oklahoma. After years of research by sources, no one can seem to recall or produce anything remotely scandalous or unusual about William. Since Cape was quite a long drive from the largest donors in the more corporate-based major cities of St. Louis (100 miles to the north) and Memphis (175 miles to the south), Reverend Huffman may well have had great difficulty in courteously roping in substantial contributions. America was still slowly rousing itself from the dreadful economic slumber of The Great Depression and not a lot of citizens or corporations in mostly average-to-low-income southern Missouri had much money to give. The specific Red Star District that William lived and worked in was decidedly middle-class but hardworking. They'd given many times in the past and were becoming tapped out, to make things tougher on Clergyman Huffman.

Quietly tenacious William would not go to work as a full-time evangelist and fund-raiser at the Red Star house of worship until that November, according to the *Missourian*. The same newspaper's edition delivered on local doorsteps the late afternoon of Saturday, April 12th, 1941, advertised that Reverend Harrison C. Crosley would conduct Easter services the next day instead at the barn-sized Red Star facility, the following morning and evening. Another visiting preacher - who was staying at a preacher's house just down the street from Pastor Huffman - was handling a special revival meeting that weekend at the tabernacle. Huffman's name was never mentioned in the many ads and articles on Christian services Red Star offered around Good Friday and Easter, 1941. Thus devout William clearly had some free time on his hands that Passover/Easter weekend, waiting for more religious and financial duties to come his way.

Thus with no sermon to deliver the next day, Reverend Huffman was able to kick back a little that Saturday evening at his simple church-rented two-story home, perhaps reading the daily paper, making fund-raising calls, writing letters, dining, and chatting with his family. For the past few weeks his older son - namesake William Guy Huffman, Jr. - and his very pregnant wife were around the house, eagerly anticipating their first child. A teenage son dwelt in the guest room. Possibly still other relations and friends visited that holiday weekend. It was likely a happy, relaxed family gathering as the evening of the 12th wore on, with thoughts of turning in within a few hours, in order to be ready to attend their church's Easter Sunday services the next morning, and perhaps the town's sunrise service, or the Masonic Temple's special program on Broadway, or even the city's Easter parade.

While today it is not one hundred percent certain that the stunning extraterrestrial events that took place near Cape Girardeau most definitely "fell" on this specific Saturday night it certainly fits with the rather thin dating evidence uncovered so far. For instance, William Huffman's granddaughter Charlette described the occasion as happening within a "three or four weeks" time-frame before the May 3rd, 1941, delivery of her older sister in a Cape Girardeau hospital. Saturday evening, April 12th, fits perfectly for this and a few other reasons that will be disseminated later herein.

Things were especially quiet on this night. The evening newspaper strewn about featured a cover photo story about a deceased senator's body being carried out of the U.S. Capitol Building, after lying in state. The nearby big brick International Shoes factory, once specializing in "Red Star Shoes" - which gave the church its name - was located just a few blocks to the south (now gone). The factory had no weekend evening shift to let out, not on a holiday. For Christians, Good Friday had passed and the various holiday Sunday activities were a good night's sleep away. Nearly everyone in town was likely quite at ease. The sun officially set at 6:33 p.m., according to the papers, and dinner was likely served all over town around that time. Perhaps within ninety minutes or so, all of that serenity changed greatly for several area citizens, in a sense for the rest of their lives.

As the sunset slowly faded and nightfall set in, grateful and gracious William and Floy, age 43, were likely fussing over their lovely daughter-in-law, now about eight and a half months "expecting." At this somewhat modest but cozy house, the circle of Huffman family and friends enjoyed a big meal and the unusually warm, light breeze seeping in their wide open windows. The pastor's two sons were perhaps listening to radio shows, playing board or card games, and rehashing old family stories. William Guy Huffman, Jr., was age 24, while single Wayne was 22, not as close as they used to be as the older brother had work and marital issues to deal with daily. His young wife probably spent time talking about childbirth and infant care with experienced Floy as they did the dishes. All was quiet, until...

Suddenly the God-fearing minister heard the call; he stood and answered an unexpected telephone's ringing, sometime between "eight to nine o'clock of the evening," it has been estimated by his granddaughter. {Another Charlette Mann account stated the time as "between nine and nine-thirty," so we'll stick on "around nine o'clock" as the general time-frame.} It was a simple yet rather urgent call to service; William was quickly informed by someone affiliated with the Cape Girardeau police department that he was needed at a sudden, tragic-sounding calamity outside of town, perhaps even just outside of the county. A terrible airplane accident! Would he go help in any way possible, mostly spiritually? It was urgent.

The general report was this: an airliner apparently didn't make the sunset deadline for landing at the unlit Cape airfield and came down in the dark not too far away. A very frightening fireball was sighted at the crash, rising from the remains on a local farm. A resulting blaze was still ongoing. Could Pastor Huffman hurry there and offer blessings, spiritual comfort, and heartfelt prayer for the victims, alive or dead, that were sure to be there?

We can estimate now that around eight o'clock to eight-thirty p.m., smoke and flames were first lighting up an obscure rural farm property, which was dotted by some trees and shrubs with new green buds. Stunned, the anxious crash eyewitness rushed for the nearest telephone. There is no guarantee this rural farmer actually owned a phone in those Depression-era days; he might have been forced to race to the nearest home or business in his community that was connected to the outside world. When he found a working telephone, the stunned farmer called the tragedy in to the nearest and largest firefighting organization he could think of, first by way of urgently contacting an operator on duty at the telephone company's switchboard building in Cape Girardeau, since there was no "direct dial" for many in 1941.

The Cape Girardeau police station on 538 Independence Street was physically shared with the Fire Department. If the Cape police were first called and notified, then they swiftly and easily spread the news to the firemen within the brick building, or vice-versa. This was necessary on substantial-sounding cases that would involve both entities; quite often the Cape police would tail the Cape fire department on calls if they were headed out to a blaze, to help quell emergencies and properly examine the situation for any possible criminality involved.

Everyone at the Cape fire/police HQ had to be on alert to respond to any emergencies, and frankly shocking, sudden news like a "plane crash" and resulting "fireball" was of great alarm and personal concern - and virtual entertainment in a quiet small town - during an otherwise uneventful evening. It could have involved local residents as victims, people that some in both first response departments knew. For both departments, the situation might have been on a quiet Saturday night a serious case of "all hands on deck" to aid not just the farmer's family and property, but the sad victims from the crashed plane, if any were left alive. This would likely have meant calling in - again, by way of a phone company operator on duty - some of the local "auxiliary police" and "auxiliary firemen" in the general area, mostly American Legion or militia-trained volunteers and retirees who were dedicated to emergency response situations. They at times worked in conjunction with the regional Red Cross, staffed also by unpaid volunteers. Any one of these people could have been contacted, readied themselves and their equipment. But this was Saturday night, with folks out having fun, away from their homes and businesses, and since the endangered address given was out of town and rather obscure, somewhere in the maze of narrow farm roads and long stretches of freshly furrowed crop-land. in the dark of night. well, getting anyone there in a timely manner was a substantial problem for all concerned.

Professional firefighters in tiny rural farm communities outside Cape Girardeau - like nearby Chaffee - were either small in number or nearly nonexistent, although loosely-organized volunteer groups were fairly common. On a Saturday night these men were likely fanned out across the community too, with almost no one left at what small, crude station houses existed. Bigger and better trained and equipped Cape Girardeau firefighting and law enforcement officers, they were the ones that needed to be contacted foremost. Along with medical first responders, coroner's office personnel, and now clergy. Oh, and the county sheriff's department (located in nearby Jackson), the local FBI office, and maybe even the newspaper's headquarters. They all scrambled with nimble dialing and helpful operator assistance to first inform each other of the news, then assessed what they'd need to take with them to a flaming plane crash site. Then they hurriedly fired up their vehicles to rush to the countryside address given. On a holiday weekend, this became a bit dragged out as precious minutes ticked away.

In 1941 Cape Girardeau, there were only fifteen total police officers, working of course in different shifts, only a few at the station at most, with some men out on patrol (no women were employed, not even as secretaries, evidently). For such patrol duties the men utilized the two or three working squad cars available, one with a brand new mobile transmitter. They also had two motorcycles. Low-level cops were out on "foot patrol" on occasion and were not available for shadowing the fire unit personnel who set off to a rural "plane crash." Some cops were considered police "clerks," manning the station, its phones, and its paperwork, while still others were established upper level "magistrates" or high-ranking officers who did not ordinarily respond to fire calls, such as the courageous police chief himself, Edward W. Barenkamp. Then there were plainclothes detectives; they might have been utilized if they were still around on a Saturday night, not always a given. A few other of the paid staff had already worked their set hours, or were about to, perhaps on the overnight "graveyard shift." So out of the fifteen members, only a few were on call that night.

At any rate, it was always a long day for each uniformed cop; a newspaper account from just days earlier mentioned how Cape Girardeau policemen worked very twelve-hour work shifts, unheard of in today's world. That's how stretched thin the force was in April of 1941. If a cop went on duty at nine a.m., he was still around to respond to calls at close to nine p.m., pretty worn out.

Meanwhile, the Cape fire department consisted of eleven men, also obviously working in long shifts too. Some toiled at the stationhouse, some dwelt quietly upstairs within the building, and a few lived in town, depending upon their family situation and rank. The city's young mayor recently boasted of obtaining for the department a brand new Dodge "ladder truck" in an article that appeared that very late afternoon, in the afternoon edition of April 12th's *The Southeast Missourian*, the long-standing local newspaper still in business to this day. The new truck promptly got dinged up that afternoon in a traffic accident, on its way to a small blaze.

In the growing darkness of the rural farmland, it was still difficult to tell what precisely had landed abruptly, really. It was simply assumed at first to be a plane, maybe even a large airliner with many needy victims - and corpses - from the description of the fireball. Lots of jet fuel must have been involved. No. upon a closer look at least some of the wreckage could be viewed in the blaze; it was a smaller aircraft, not much to it, really. Oddly rounded. The wings must have fallen off it. Maybe only a few on board. After making the call the frantic property owner likely kept any curious children and pets at bay as the family hunted for any flashlights or lanterns, first aid kits, and even buckets, blankets, and water, to take to the burning site outside. Possibly a portable extinguisher was available. Concerned neighbors might have begun to arrive on foot, as well, having heard and/or seen the frightful crash. The smoke and flames spread. Containers of water, from a house, a well, or a pump were probably ordered up. Shouts and chaos in the dark.

Just about nine p.m. a Cape policeman (or police associate) was calling William to ask if he could come out and pray at the site. Perhaps lend spiritual comfort to anyone who was needing medical and emotional assistance? It seemed like there could well be upset, traumatized survivors with friends or relatives who didn't make it, gruesome casualties in a somber-sounding scenario described insistently over the phone to Cape headquarters. Some might require the "last rites," ritual prayers issued over the dead, which admittedly is a Catholic policy, not Baptist.

Apparently Reverend Huffman never named the police source who contacted him that night, nor which one came out to his house to collect him. Could they have been the same man? *The main question is: how did any law enforcer decide to call seemingly obscure William, of all the many Christian ministers in Cape Girardeau?* Either way, Minister Huffman answered the phone, then answered the call to service. But precisely *who* placed that call?

To the Cape police, this "plane crash" was serious business and perhaps a growing news story with national implications. No one wanted to screw this up from a medical, police, political, or legal standpoint. If a Cape policeman or someone associated with the department was idling in the station house, or within an easy phone call's reach that night, they were likely pressed into duty. Anyone of city or even county importance would logically have been urgently requested - or *ordered* - to hurry to the scene – if they were home. Factoring into this was the real possibility that the messy, unguarded crash site might be looted or altered by someone with less stellar moral standards, operating in a rural, unpoliced setting. Criminal conduct at the accident scene was always a possibility; cops were needed at once to keep the peace and help aid the injured. To protect and serve - *and* record the facts quickly and efficiently. The sticky situation could not be allowed to be exploited or legally botched and end up embarrassing the whole department, the city attorney's office, and the community as a whole.

Included in this emergency response scenario might well have been Cape Girardeau's resident Justice of the Peace, Milton Cobb. If Milton had been at CGPD headquarters on Independence he would naturally have been pressed into duty. If not, he might well have been called at his home. An experienced and trusted justice like Mr. Cobb was known in many communities as a "police justice" or magistrate who worked almost daily with the local police department in resolving legal matters, including challenged traffic and domestic dispute situations. He might well have been with the aforementioned "police auxiliary group" that backed up Cape cops in emergencies. As part of his duty in Cape Girardeau County, Milton's official presence and signature might have been required on death certificates and legal records documenting any gruesome "airplane crash" scene. Injured bodies, both dead and alive, seemed certain to be on the ground, and the legal "chain of evidence" had to be recorded for any possible pursuant criminal investigation or inquest, and later trial. There needed to be a legally trusted eyewitness that there was no looting or messing with the victims by anyone, nor with the damaged airplane and its scattered luggage at the chaotic scene. Testimony would be required later if sabotage or murder was found involved in the plane's downing, within the subsequent inspection and investigation by authorities. This was a very serious, possibly high profile legal case. Hence Milton Cobb was very likely called upon to help that hectic night. If he was relaxing at his Cape Girardeau house at the time, he was probably dialed up within the first few minutes and informed he was officially and legally bound to respond to the seeming medical and legal emergency at hand. *Get dressed, you're going.*

Yes, Milton Cobb at age 47 was more than likely a real "player" that April 12th *and* a key figure for us in understanding who specifically reigned in Reverend W. G. Huffman's assistance. Why? *Because Milton lived directly across the street from William.* Their houses stared down at each other. They *had* to have known each other from their down time in the previous months, puttering around their yards and tending to chores as they chatted that warm Spring, while other neighbors often worked nearby shoe factory shifts. In those days, neighbors commonly socialized and spent much time in conversation, since there was no television or internet, or other indoor distractions to keep them from outwardly interacting often with others. When these two local men talked they discovered they were separated in age by just a few years, were both happily wed, and that both had experience in marrying couples and in counseling and settling disputes within area families. In fact, the two neighbors were so tangibly close that if called by cops at home that night, Justice Cobb simply had to peer out his window and notice that across the street the good pastor's lights were on and people were moving about inside his cozy, nearly identical white-painted home. As he learned of the tragic-sounding emergency, the thought likely occurred to Milton to import the friendly, calming pastor. Godly prayers. Perhaps last rites. Spiritual aid and comfort. It was just as important to some in those days - in a deeply Christian community - as medical care or firefighting efforts. It's possible both Red Star District gentlemen were very devoted Southern Baptists who felt God guided their lives and wanted them to helpfully get involved in a crisis response like an airplane crash. It was much the same Christian attitude shared by the Episcopalian U.S. President, Franklin Delano Roosevelt, age 59, spending that particular quiet evening with no family but a few guests – like an Army officer, and aide Harry Hopkins - at the White House, planning on attending early morning Knights Templar ceremonies early the next day, and Easter church services in Washington D.C. right afterwards.

If previously relaxed at his North Main Street abode, alert Milton grabbed his phone and a city directory, or a simple notepad of scribbled numbers of friends. Justice Cobb could actually *see* the preacher near his own open, bright, living room window, interacting with company across the street, all as Milton dialed. The justice probably knew from experience that William was a healthy, caring, able evangelist, and would lend a hand if asked.

Charlette Mann recalled that the family history described someone associated with the Cape police force calling the Huffman home, and later a police source *not in uniform* soon arriving at the house in an unmarked car to take William to the "plane crash" in the countryside. Such a depiction neatly fits police Justice Cobb, wearing casual attire, pulling his own automobile out of his very narrow gravel driveway across the street and guiding it into the fund-raising preacher's driveway in a scant matter of minutes. Milton never wore a uniform, nor drove a squad car. He was more like a "police associate," working with the Cape department on various matters at the least on a weekly basis. He's the logical candidate, but admittedly there are others, off-duty officers who would have been called in and perhaps thought of kindly W. G. Huffman on Main Street, and yet again he was awfully obscure and unknown to most.

Back at the Cape police/fire station, the policemen in charge undoubtedly called a local hospital. They would need to order up an ambulance team of medics to rush to the crash site, either through the hospital itself, or a singular ambulance service in town. He might have even called a mortuary, for hearses were also often utilized in lifesaving roles back in the day. The law enforcement, firemen, and medical personnel all would likely have welcomed an extra hand like the justice and his trusted minister, if not just to help lug bodies to emergency vehicles or pull back crushed fuselage to allow crash victims to escape the flames. It was time to *minister* to the less fortunate, for sure.

Since William G. Huffman was not busy that Saturday night, and other Red Star pastors clearly were, he knew he had to attend to this pressing matter when asked. Going to a plane crash site would actually have been pretty important to the good reverend, for several reasons. From a humanitarian perspective, and for spiritual reasons, sure, but it might also mean future donations to the church, and perhaps even newspaper headlines and good publicity for the entire Red Star Tabernacle community. And that was what William was imported to do in the first place. Perhaps he even added to the policeman on the telephone, "I'm not Catholic, but yes, I can handle "last rites," if it is requested by anyone there," but this is purely speculative.

Inside his cozy rental home, worried W. G. Huffman picked up his favorite cloth hat and his Bible, and likely said a little prayer for the strength to get through what sounded like an ordeal to come. The more he thought about it the rougher it seemed like this was going to be.

There might be agonized burn victims in very rough shape, ghastly corpses in various states of dismemberment, and stunned survivors that were in pieces, literally. Maybe burned and/or bloody body parts on the ground, amidst chaotic private possessions and luggage strewn around. Twisted metal and jet fuel blazing. What a horrible sight and night this was going to be. A real life nightmare, perhaps.

So this is where Reverend Huffman's *Alice In Wonderland* adventure began. Since it was an unseasonably warm night, William might have grabbed only a light jacket with his hat, but even that was unnecessary. Records show that the 1941 Easter weekend in Cape Girardeau was particularly *hot* during the day, with highs in the eighties. It was likely in the upper sixties to low seventies outside at the moment, the sunset's afterglow also now gone. Therefore perhaps William only clutched his Holy Bible and headed out the front door. Once outside, he hurried down the porch steps and across the yard to meet the fast-moving lawman who arrived with a honk in just minutes.

The two men probably departed the Huffman home around nine to nine-fifteen p.m. or so. The Cape Girardeau police/fire department headquarters were located only a few miles and minutes from William's home, if that was the original starting point for the trip for the police associate. Even less if he came from across the street. The unusual duo undoubtedly rode off at a fast pace, headed south down Main to Broadway, then out west to Highway 61. From there it wasn't the easiest drive, what with very few street lights over often-unpaved, narrow country roads at times, as urban Cape Girardeau gave way to dark, undeveloped rural lots and farm fields, lined by thickets and woods. They took perhaps up to thirty minutes to get to the proper agribusiness area. The further out they drove, the worse the driving conditions became.

The anxious companions unwittingly crept closer to genuine history in the making, the most astounding sight of their entire lives. There, up ahead, *finally*, they could see the glowing lights of a rather strange backyard fire, near a dirt field and farmhouse. They observed some vehicles parked on the side of the road, in the grass and weeds. At least one fire truck, one cop car near the farm, and perhaps an ambulance or hearse. and plenty of other vehicles were parked in a haphazard manner by the street. *This must be it.* The twosome witnessed silhouettes in the moonlight heading to the site, or some just leaving, numb. William's driver wedged their vehicle in a grassy parking spot by the side of the road and the two got out, full of anticipation.

All of this activity took place fairly innocently, with no hint of otherworldly overtones at first, according to William's wife, Floy Huffman, who was dying of cancer in the early 1980s. Floy was staying at that time with one of her two granddaughters, Charlette Huffman Mann, as a guest at her home in Tyler, Texas. That's the town where Charlette worked as a spiritual counselor at her local church. "I am not a UFO person," Mrs. Mann admitted from the start, perhaps bolstering the credibility of her case. Think of Charlette and her family tale what you will, no skeptic or investigator has uncovered any deceit in her eye-popping, unchanging claim or in her personal life.

Weak but lucid, long-widowed Floy Peters Huffman was largely bedridden as cancer ate away at her health. A tough Christian soldier, Floy endured grim chemotherapy and radiation therapy in 1983 and '84 as bravely as possible, realizing her time to shuffle off her mortal coil fast approaching. Charlette had heard the family's whispered ET rumors and seen a related small photograph many times, decades earlier (see Chapter Two), but didn't know the whole saga, front to back. She knew only tantalizing bits and pieces. It had been eating away at her for years.

She knew that her Gramma Floy needed to spill the beans soon on what Grampa William had gone through that strange springtime evening. It was urgent, Charlette thought, that Floy soon tell every detail possible, so she kept at it, peppering her grandmother with careful open-ended questions until she finally broke the beloved octogenarian woman's vow of silence and talked at last.

Thanks to Charlette repeatedly interviewing fading Floy and taking copious notes, we have a pretty firm understanding today of the most amazing extraterrestrial encounter of all time - at least it was until that point in history, April 1941. Japanese forces were running roughshod over Pacific Rim nations while Adolph Hitler's Nazi Germany was taking over much of Europe and secretly planning to invade Russia in about two months' time. In quieter, pre-war America, things were only slowly heating up towards an eventual showdown with the Axis Powers in World War II, with neutrality and peace largely the mind-set of the day, along with preparation "just in case."

In her 1984, essentially-deathbed confession, ill Floy Huffman at age 87 explained that her late husband William did not return from his unexpected 1941 jaunt out with the friendly police source until "late at night," perhaps around midnight, and when he did, he was a changed man. Shaken, pale, hushed, and somewhat frightened, William would never be quite the same again, becoming much quieter, more introspective, and yet more tolerant of all the possibilities in life, Floy assured Charlette. William made an impulsive decision: to tell all in the house what went on that fateful night, and in a sense when he did his wife and two sons would not be the same, either. When a still-staggered W. G. Huffman gathered the family together in the small living room of their modest Cape rental home, his concerned sons readily stepped up to hear the tale. The two twenty-something men were anxiously awaiting their mild-mannered father's return from "the airplane crash" and wanted full gory details.

"I'll tell you this once," emotional William blurted out, "and then I'm never going to speak of it again," Floy recalled him declaring, and then adding, "In fact, *you* can never repeat it, either." *That* got their attention fast. All sat on the simple furniture and fell silent for the disturbed reverend to continue. By the time he was finished, the devoted wife and two caring sons may well have been on the edge of their seats. It was not the kind of tale conservative Christians in a small town told in public, not if they wished to keep their jobs and respected positions in a rather button-down society and era. Here's what the good pastor spilled that spooky midnight...

First off, William said, after getting out of the parked police friend's car in a grassy lot by the farm road, the two men hurried on foot to the crash site, featuring reddish-yellow flames, still burning in scattered places, and the locals' various search lights, creating a substantial, moving glow in the night sky. At perhaps around 9:30 to 9:45 p.m. the duo walked across farmland "about a fourth of a mile," William later estimated, past trees and shrubs which were just starting to turn green, to get to the weird crash scene. It wasn't quite the growing season, so the moonlit nearby fields were simply furrowed with fields of upturned grayish-brown soil, mostly without even pale sprouts yet. Grassy pastures and farm yards were likely ankle deep, however, unlikely to have been mowed yet.

A flame-ridden crash site up ahead came into clear focus as the duo plodded around a barn and farmhouse, towards a small crowd. Excited voices chatted away. Shadowy, stunned adults moved about, and perhaps a few children and pet dogs wandered around curiously, too, all slack-jawed and a bit uneasy. Some were simple farmers. Others were concerned neighbors, all dressed informally, even shabbily. Yet educated, professional men, a few in uniforms, some in suits, many with duties to perform, they dominated the immediate area, milling about. A fascinated "man from the newspaper was there," probably *two*, William remembered, taking bright flash photos now and then, adding to the wild light show. Some of the men were in dark suits or shirtsleeves, combing the site, taking notes. Flashlight beams and lanterns and some vehicular headlights doubtlessly helped stab the darkness, making it easier for William and his friend to make their way past a few healthy trees to the wreckage. They likely had to step over or around a fire truck hose, farm implements, and pools of water here and there. Perspiration was growing on the brows of perhaps as many as a dozen or more white males who sifted through an odd debris field in the fairly warm air, their faces illuminated by the flickering lights. The atmosphere was surprisingly *not* urgent, but surely stunned.

Airplane accidents in those days were not that uncommon. Flight technology, air safety regulations, and knowledge of weather conditions were certainly not what they are now. Yet as William looked around he did not see any human bodies. There was only the strange, grayish-silver, metallic debris scattered here and there, glinting in the lights, and the remaining little flames that were being carefully smothered. The Cape fire department was doing their best to inspect the situation and put out the blazes, but most of all, they were curiously staring into the fuselage of the odd craft that had so recently crash-landed. And at its lifeless, small passengers in the fragment-filled, scorched grass. *Finally, a victim.* Then William noticed a second, and a third. Heads, torsos, arms, and legs... all attached and intact, not even burned or cut! They seemed wrinkled and human, like three thin, big-headed children, yet. *rather strange.* The bodies were somewhat *gray* in color, long-limbed and rubbery-looking. People stood in a daze and stared down at the lifeless strangers, quite perplexed. Something was very wrong here, something *very* peculiar indeed...

It is possible an agent or two from the Federal Bureau of Investigation was present at the bewildering accident site, sifting through the rubble and flames. It would have been difficult for such FBI men to have been identified, since these "feds" or "G men" simply wore suits and ties, standard apparel for agents under strict, conservative director J. Edgar Hoover's control. Unless of course William Huffman heard the man audibly describe himself as an agent and/or he flashed a badge. The city of Cape Girardeau had just opened its first FBI field office one month previously, conveniently. Arlin E. Jones, the resident agent in charge - if he was at home that night - was likely called in by the police. Jones lived with his wife on Broadway in Cape and worked down that same street, and could easily have been rushed to the scene by a friendly cop since he was new to town, having just been hired. Otherwise, if Arlin was on his own, finding his way through town and then the maze of country farm roads in the dark would have been a daunting task for a newbie. He likely arrived pretty late, if at all that night.

One thing was for certain: the entire affair needed to be carefully investigated. What had taken place here, something criminal? Deliberate sabotage? On March 12th, 1941, a local newspaper article mentioned how the FBI was already in the area, undertaking "surveillance" on possible un-American "subversives," suspected spies and saboteurs, and general criminal conduct. And praising local law enforcement for their assistance in coordinating information and operations. So a call from the cooperative Cape cops that night to Agent Jones is very possible, nearly a sure thing.

Some of the men present were described by Huffman later as "uniformed" while others wore simple suits, their colors undoubtedly gray, black, or brown. Hats were likely either felt-type fedoras or wide-brimmed style, as was the style at the time for so many men in American society. Loose, almost baggy slacks, white long-sleeved shirts, and a few wire-rimmed spectacles. Close-cropped haircuts and little in the way of jewelry or tattoos. Cape Girardeau police uniforms were crisp and military-like, featuring dark blue blazer jackets with gray slacks, topped off by off-white caps with shiny navy blue bills. Firemen uniforms were red and gray, and looser, for easier movement in fighting blazes. Volunteer city police and/or firefighters - if any - simply showed up in whatever they may have been wearing at the time, of course. County sheriff's office representatives wore lighter gray uniforms. Army personnel wore khaki or tan-colored uniforms. Women in those days mostly wore loose-fitting dresses and hosiery, hats and even gloves when going out; slacks for any adult female in a conservative town in 1941 were almost unheard of. Children generally wore whatever they were told, not whatever they pleased.

It wasn't *just* fire personnel and law enforcement men in uniform, Charlette remembered of what her grandmother told her, there were also *a few uniformed military men* in the murmuring crowd. Without much doubt their faces and reactions were also blank, puzzled, riveted, and perhaps even a little pale with fear, staring down and around at the crash field's contents. This was not like any airplane or blimp or balloon they'd ever seen or heard about, this crashed metallic mess on the ground. It was growing obvious that it was *otherworldly*. The reality sank in as the minutes passed and the whisper grew into audible gossip.

There were three "people" present at the site that may have worn the oddest attire of all: wrinkly, tight-fitting, silver-gray flight suits, appearing almost like crinkled aluminum foil. Or was that just their crinkly, trauma-and-atmosphere-affected *skin*? William was unsure. These odd, silent figures seemed - all stretched out - to be about only four feet in height, and remarkably thin. "They had been taken from the crash," was how Charlette recalled of her grandfather articulating to his family. Physically removed, but too late. Only one of these odd entities even *moved* when the shaken pastor walked around, staring at the bizarre scene. Two were silent and motionless, seemingly dead as they lay on their backs in the grass, side by side, close to each other. "They didn't appear to have any injuries," was what puzzled W. G. Huffman mentioned to his amazed family that night. *Intact "little people" from another world!* Lord have mercy! What an incredible find!

In hindsight it seems apparent the skinny, unearthly trio were helped or pulled gently across the soft grass by some helpful first responders, maybe even the property owner himself, likely the first person on the scene. Recall the term *"taken from the crash."* Likely physically *removed* from the opened vehicle, by *someone*. At any rate, the odd bodies were assembled rather neatly yards away from the eye-catching vessel, sometime before William arrived with his incredulous friend. The broken-open aircraft that "had no edges or seams," Charlette said grandfather asserted, became quite the curiosity. William never used the term "UFO;" it simply hadn't been created yet. The weird beings before him were "little people" or "creatures," William recounted, not at any time using the term "aliens" or "extraterrestrials." These descriptions became commonplace only decades later.

Reverend Huffman wandered about the crowd and gazed at the scene. This downed "ship," this unusually round and amazingly smooth *space*ship, was now in various sized chunks, one piece noticeably much larger than the rest.

The other smaller pieces of the metallic chassis, and many, many tiny shards from the explosive impact, littered the surrounding flame-filled debris field. The unusual craft had obviously suffered a dreadful accident, and was irreparably damaged. It could not possibly just reset and take off again, even if the crew were alive. It had no exterior numbers, pictures, or markings, with no engine visible. No wings, struts, or propellers, nor exhaust pipes or vents. In retrospect perhaps the vehicle's external chassis had the appearance of perhaps "brushed nickel" as it is called nowadays. *It was unlike anything man ever made.*

One could not help but notice at the scene that there were no luggage or purses on the ground, no paperwork or flight plans, no pilot uniforms or airliner accouterments. There was no sign of *human* life anywhere within the flaming debris field, meaning no airplane was involved. There was no collision debris of any kind, no evidence of manmade involvement at all. The sight of it was stunning, perhaps more than some could absorb. But since there were also no reported *alien* material goods present - extra flight suits; food or drink; containers or messages; or a bunk-filled "sleep chamber" within the craft - then it would seem a pretty good bet that this was a short-range, exploratory reconnaissance vessel, perhaps on its way from or back to its larger mother ship. Upright seats only, with no other room, evidently. A short-range "scout ship" or observation "shuttle craft," as it were.

For William, it was pretty clear that the two speechless, inert "Grays" – as we say today - were quite dead, and another ET a few yards away was very close to it. All three appearing the same. And these were the specific cause of his need to be there. He had to focus now. To attend to the victims' suffering, be it physical, emotional, or spiritual. To care for the uncared for. Yet like everyone else,

William stepped in closer to examine a lithe and creepy "man." How is it they all had such identical big heads and black eyes!

The scattered fire was just about out now, smoke wisping up, a crisp burning smell likely filling the air. So did excited chatter and hushed guesswork. Who *were* these "people"? How could they look so exactly alike? Where in the world did they come from? William uneasily thumbed his Bible as he stared down. What in God's name was going on? These nonhuman beings looked like little demons! Were they somehow "fallen angels," or creatures from hell? *Was this the start of the end of the world?*

Clergyman Huffman did not recall any of the three alien bodies spilling any blood or oozing any fluids, but there would have been some amount of danger present, especially if the evangelist had handled them (which he said he did not, wisely). The possibility of radioactive chemicals or unfamiliar otherworldly germs and toxic bacteria doesn't seem to have been in play here. Perhaps it was perfectly safe, for whatever reasons. Maybe what dangerous bacteria was initially present burned up in the entry into earth's lower atmosphere. But even the affected farm site itself *could* have been troublesome for anyone without a "haz-mat suit," which simply did not exist in those days. Perhaps few present gave any heed to the danger of actually touching a non-terrestrial craft or body, and thus opening themselves up to viral infection and illness from traces of foreign, deadly chemicals, or poisonous bacterium of various unknown strains. If anyone got sick afterwards, it is not recorded, but would not be surprising in retrospect.

The Fire Chief of Cape Girardeau, Carl J. Lewis, was probably notified immediately and likely rushed to the scene that night. If word had spread of an airplane crash creating a fireball, with resulting flames and emergency services required, Lewis would almost certainly have been a galvanizing force for his squad to gather and rush to the site in the most timely and professional manner possible, even if called at home. Local volunteer firefighters might also have been assembled and present, or at least on their way. Fire trucks of that era featured a modest water tank on board, one that needed rubber hoses and noisy pumps to get a strong flow going on a blaze. Out in the country, there would be no water hydrants to tap into, but perhaps a well. Portable extinguishers and thick, smothering blankets were on the truck and doubtlessly utilized too. If firemen *were* at the scene as claimed, they were the prime candidates for having removed the victims from the wreckage, an unpleasant part of their laborious jobs normally.

Strangely enough, fire was reported all around the crashed craft and its occupants, but W. G. Huffman claimed *he did not recall seeing a single scorch mark or fire damage on the ship, the debris, or on the diminutive ET bodies themselves.* This is a small but corroborative clue for a few later stories of recovered alien bodies supposedly in America's secret military control.

Policeman in plain clothes were also supposedly there, roaming around, Huffman said, probably helping to stamp out flames and scribbling notes regarding the unusual circumstances. These were most likely some on-duty law enforcement officers and/or police detectives, maybe having called in a few off-duty pals. It is reasonable to speculate that on a Saturday night, one particular off-duty Cape Girardeau cop in "civvies" who was contacted and rushed to the site would have been tough Chief Barenkamp, a man so active, unafraid, and aggressive he would leave the force to join the Navy "Seabees" later that year.

It is also very probable that someone alerted the Cape Girardeau County Sheriff, Ruben R. Schade (pronounced

"Shoddy" locally). One of his surviving brothers gave researchers decades later a clue that Ruben was likely alerted. Clarence Schade, just 26 at the time of the crash - told an investigator in the 1990s: "Yes, I heard of a spaceship with "little people" back a long time ago, but I didn't believe it." Even though that was all elderly Clarence had to say about it, it was a solid Missouri crash confirmation, of sorts. Obviously Clarence wasn't called to the scene that night, but someone he knew - someone *quite in the know* - was there and willing to break a vow of silence on the matter to take young Mr. Schade into his confidence.

Sheriff Schade was nearly a week away from his 36th birthday when the big event occurred, but in all his days he'd never seen anything like this mess. Ruben had spent the previous few years living and working in Jackson and Sikeston, Missouri, as a respected newspaperman. He had personally reported on some local stories as they unfolded. A conservative "law and order" type, Ruben was elected the Cape County sheriff in early November of 1940. Another Schade brother, Ben, age 31, was a Parks Air College "purchasing agent" at the Missouri Institute of Aeronautics, the respected Army pilot training program at the Sikeston airfield, around thirty miles south of the crash locale.

Since urgent news of an "airplane crash" was called in to authorities, it is certainly possible the property owner directly called the Cape County Sheriff's department, and they in turn quickly notified Ruben Schade. Or, when the Cape police/fire department got the startling information, they felt it was an "all hands on deck" situation that necessitated them calling Sheriff Schade's office, in nearby Jackson, Missouri. From there, Ruben might have called trusted, mature Ben, since his specialty was Army airplane training that could have been the cause of the accident in the first place. In that early innocence about the nature of the crash, both Ruben and Ben might have felt they either might know the local airplane pilot and perhaps his passengers, or that it was an errant MIA flight. Either way, they'd both want to go immediately to the site and investigate, for various reasons.

Seeing as how they were a fairly tight family, the married Ben and his wife might well have been spending the Easter holiday weekend with his wife at home, or with his parents, or his wife's parents, living in the Jackson/Cape area. They might even have been with Ruben and his fiancé at the time of the call. It is easy to conceive of Ben getting a quick patrol car ride, perhaps even from Ruben himself, to the rural crash site.

Floy Huffman evidently told Charlette Mann that William reported seeing "some men in military uniforms" in the crowd when the pastor first arrived at the crash site that night. While this could have been anyone, remaining unidentified, it would make sense for Ruben to have imported Ben and perhaps any MIA representatives he was with at the time. From there, one of the U.S. Army personnel present at the site - Ben himself? - contacted their Sikeston flight training colleagues. By the time startled Preacher Huffman walked around the distinct debris field, the Army airport in Sikeston was undoubtedly buzzing with activity. In fact, a contingent of trusted men, likely in an ambulance-type vehicle (with the red cross painted on the side), was already speeding north, to the accident location. A limited number of cadets and their superiors had scrambled to assemble vehicles, helmets, crates, body bags, and probably guns and ammo. It was for those present a real "red alert." Was it just a drill? Was it quite the opposite, the start of a deadly serious interplanetary war? No one knew quite what to expect as they hurried about in their barracks and out on the tarmac. But it is certain that the Army reps from Sikeston knew it was an alien crash, not an airplane crash; they came with a show of force to prove it a few minutes later.

At the time of the bizarre air accident, Sheriff Schade was engaged to be married. About two weeks after the crash, the local newspaper reported a Schade family reunion, likely to discuss the coming June wedding's many details. That may not have been all that they talked about. Ben was said to be there, Clarence undoubtedly too, while still another Schade brother suddenly departed the get-together for a trip to St. Louis, in order to join the Army Air Corps. Ben, Clarence, and the younger brother - all now deceased - also attended Ruben's marriage ceremony in late June, according to *The Southeast Missourian*. It would have been at one of these two family meetings, one could easily surmise, that Ruben and/or Ben attempted to tell Clarence Schade about the secret goings-on regarding the odd crash outside of town. The "spaceship and little people." But since the two brothers could produce no real proof, and it all sounded so ridiculous – amidst possible alcoholic beverages served - no one believed the Schades in the know. They likely seldom if ever discussed it again, not wishing to appear unstable or untruthful.

Ruben Reinhold Schade left his esteemed county office in 1944 and spent the next three decades as a conservative reporter/columnist for the *Missourian* newspaper, evidently keeping pretty mum about the crash. If his deputies arrived with their boss that night, or separately, it was not specifically noted by Reverend Huffman. *However, William was not the first at the scene, so many people could have come and gone before he arrived.* Although admittedly it would have been very difficult to assess *this* mind-numbing scene and then simply walk away. And it may also well be that the sheriff of another, nearby Missouri county might also have been notified back then and rushed to the scene, also with sworn deputies by his side.

Strangely enough, the top candidate for the second sheriff present at MO41 also left office in 1944, to spend several decades quietly living in the community, along with his brothers (and sisters). Since some feel strongly that the crash occurred in a field not too far from Benton, Missouri, a sleepy small town in neighboring Scott County, Sheriff John Hobbs might well have been present. Benton was the county seat, home to the sheriff's office, so Hobbs was possibly in the general area that evening. It is even possible he was in his office at the time, but if it truly was a warm Saturday night, a loyal deputy was likely on duty while Sheriff Hobbs was elsewhere. But for something of this magnitude, Hobbs was probably notified wherever he was situated, and could have swiftly sped to the scene since newspaper accounts say he spent some of his free time with relatives and friends in Benton and the small town of Chaffee, not too far away, to the south. Chaffee featured only a few thousand residents, and was still licking its wounds from a May 1940 tornado that tore through town, requiring a Cape Girardeau police response. And then there was Sikeston, the largest city in Scott County.

Sikeston, Missouri, sported a population of about 7,000 people in 1941, situated some forty miles south of Cape Girardeau, on a very flat plain. Thus the Sikeston area was perfect for landing or launching airplanes needing long flat runways. The Missouri Institute of Aeronautics pilot training program at the airfield was quite a success, doubling its number of trained pilots and the size of its campus in one year, by September 1941, according to an article in *The Sikeston Herald* (former employer of Sheriff Schade). This extra construction at the institute built up new roads and sidewalks, tennis courts, medical office, and more. The strange and rather suspicious factor in all of this is that an armed guard was eventually noted by the town's newspaper in July of '41, stationed carefully at the entrance to the simple pilot training establishment, stopping all vehicles that entered the grounds, near the large hangars and offices. Unusually, each visitor was questioned as to the nature of his presence, it was pointed out, and permission strangely had to be sought at times from the officers at the institute just to enter the campus. All this sudden security for mere road and building construction? And when America was not at war and no sabotage or espionage was reported in the Sikeston area? What precious objects could have been stored, or records filed, at the aerial training facility that would have required gun-toting military personnel to intensely quiz all who approached?

A Sikeston native many years later decided to do some digging. Former childhood resident Linda L. Wallace wanted the lowdown on the mysterious 1941 crash incident since her own father was a "line maintenance" member of the Sikeston-based Missouri Institute of Aeronautics, operating out of the local airport. The MIA was largely run by personnel stemming from Parks Air College, an affiliated civilian pilot training program that was headquartered at a Cahokia, Illinois, air field. This simple flight learning center near the Mississippi River, a quick drive across to looming St. Louis, was where the school's well-connected founder, Oliver L. Parks, kept an office. Aviation-obsessed Mr. Parks trained various civilians from all walks of life - like Ben B. Schade and Linda's father - at the Cahokia PAC and then funneled many of them to the Army flight training center in Sikeston, 130 or so miles to the south. St. Louisan Parks would occasionally fly down to Sikeston to routinely inspect PAC and MIA operations there, and may well have done so swiftly when the surreal crash incident was reported (see Chapter Four). Interestingly, perhaps tellingly, Ben Schade returned to

Oliver Parks' successful PAC primary training operation in Cahokia in the aftermath of the 1941 Missouri incident. Respected Ben then spent the next few years there at PAC headquarters with Mr. Parks until the airport operation was sold in the late 1940s, the men were that close. Did the twosome bond over their mutual insider knowledge of the incredible ET event?

While toiling on some simple family genealogical research, Linda discovered familiar names in an early presentation of the MO41 crash investigation by researcher/author Ryan S. Wood. Fascinated by the entire Charlette Mann saga recounted online by Ryan around the turn of the century, Linda investigated a possible connection between an enigmatic-sounding Missouri crash incident mentioned by a relative or two, and Sikeston's airfield MIA program, where her father toiled for a few years. Just after the unusual event, her civilian dad was suddenly switched at the PAC/MIA facility to the military, before being mysteriously transferred to Chicago and later to an important airbase in Ohio that has been alleged by many to hold crashed alien technology. It seemed like the enigmatic UFO accident may not have left much of an impact on the Missouri soil, but it did leave a big one on the Wallace family patriarch.

To be clear, Linda Wallace did not actually recall her dad passing along any alien crash secrets when she was a child and unfortunately not as an adult, either. However, Linda's kin had heard some whispers, some enticing rumors, and witnessed a few intriguing and suspicious things over the years (again, see Chapter Four). Linda raked it all in, then decided to go inspect the official records of Sikeston public entities. She found a dramatic result: *nothing. Nowhere.* No records exist at all from April 1941 via the Sikeston fire department, police department, and local sheriff's office. Further, there were no articles about any crash appearing in

The Sikeston Herald newspaper files (some on microfilm) from the Spring of 1941, although a few articles were strangely cut up or redacted (blacked out). And no records could be found from the PAC/MIA years at the Sikeston airport, either. It seemed like the end of the line.

Finally, disappointed Linda hit upon a few people with some scant but significant insight. One new source she learned of had previously spoken to others, allegedly, about "little people" that abruptly died "in an air accident," and that they were whisked away from the region's crash site north of Sikeston. Another source to potentially tap was an elderly gentleman - a former MIA "associate" - who was said to be once part of two key Parks Air College facilities, and known to her father. This man - whom we'll call "Ray" - allegedly told friends some years later after the fact that *he had been brought in to Sikeston from the Cahokia PAC airfield to collect the corpses "of crash victims" recovered and held at the Sikeston base,* and that they were not *human* bodies, either. A year after this shocking transportation process, Ray was transferred to the Sikeston MIA facility. Linda Wallace discovered that when Ray blabbed later on to friends of this long-ago extraordinary secret UFO crash event, some thought of him as "crazy." He had insisted to others that "there were other {alien} crashes not reported," referring apparently to the now-famous 1947 Roswell incident becoming well known in the 1980s and beyond, part of the national lexicon for UFO incidents. Fascinated Linda wanted clear answers directly from Ray, keeping an open mind. She set out to find him and get his account on the record. When Linda found Ray, he suddenly and suspiciously clammed up on the subject of any extraterrestrial crash incident, the interview abruptly over.

Foiled and flustered, researcher Linda L. Wallace received no further information from Ray and he died not long thereafter. After Ray's passing, Linda learned through yet another source that yes, Ray *had* spoken previously about the alien crash victims he handled. Yet thanks to the age of the story and lack of firm details, the trail as it so often does went somewhat cold. Still, dedicated sleuth Wallace "picks up bits and pieces" of the odd affair now and then, and although seemingly no one person has any great details or hard proof of the MIA's involvement in MO41, or other Sikeston-based angles, Linda produced enough tantalizing information to create her own intriguing e-book (again, see Chapter Four).

By late April, fifty-five Army cadets had finished their ten week flight training course at the Sikeston airport and were ready to be shipped out to an airbase in Texas. Even their commanding officer would be sent to this site, leaving few at the Sikeston PAC/MIA set-up with great local knowledge of the UFO accident. The pilots would later be scattered to various airbases around the country, then sent to often-deadly war zones after December 7th, so details of their participation in the otherworldly crash recovery got lost in the swirl of confused activity that became the tumultuous 1940s. Yes, by late April many of the graduating cadets took time off, and the base needed to be cleared for the next batch of students, starting on May 2nd, 1941. Yet the spotlight of suspicion is definitely shining to this day on the courageous young military men who once staffed - for ten weeks, at least - the long-closed Missouri Institute of Aeronautics at the still-thriving Sikeston airport. {Another clue: Reverend Huffman did not recall the soldiers as being particularly tall or muscular or bullying; it is interesting to note that young Army fliers of that era were all were all fairly small Caucasian men physically as it was necessary to recruit and develop such shorter, thinner types to fit inside small airplane cockpits and gun turrets of the day.}

Technically, by tradition, a county sheriff is considered the person in command of emergency militia in his jurisdiction, so while Cape County Sheriff Reuben Reinhold Schade was not in command down in Scott County, his brother officer - Sheriff Hobbs - would have had ultimate authority if he had chosen to exercise it at any crash site the Army attempted to take over. Yet if the accident location was just over the county line, which was not well marked, then Ruben would more than likely have been present and influential, but not technically in charge. And Rueben Schade reported - as did all Missouri sheriffs - to the governor, Forrest C. Donnell. Ruben and Forrest were fairly well acquainted, both being law-and-order Republicans and active Freemasons (Donnell for sure, Schade suspected). {The secret society of masonic brotherhood had great influence in Cape and in D.C., especially within the Roosevelt administration.} For something of the incredible magnitude as alien creatures crash-landing on Missouri soil, one of Sheriff Schade's first acts at the scene might have been to contact - perhaps by patrol car radio - his superior, the office of the governor. This was an emergency situation, unique in human history. Within an hour of the crash response, Governor Donnell might well have been fairly well informed, although this is still speculative.

Cape's Police Chief Barenkamp and ex-chief, now-Sergeant Edgar Hirsch - along with perhaps two of their top, trusted men, officers Marshall Morton and Frederick "Fritz" Schneider - would probably have felt it was part obligation, part natural curiosity, to have hurried to the serious-sounding accident site, even if they were not on duty at the time of the urgent call. Hirsch had always been interested in flight, and helped build the first homemade airplane to fly over the city, some fifteen years earlier. Frankly the crash was something interesting to do and see on a Saturday night (most likely), even if it was just outside

Cape's law enforcement territory. Imagine the dire repercussions if these sworn peace officers did not take action and the disaster grew to affect more local citizens and plane crash victims negatively. Inaction might have cost the cops their jobs and the Cape police department its integrity.

Marshall Morton and Fritz Schneider were at times paired as a team, and might have been called upon as such that Saturday evening. Remarkably, the twosome would go on to land coveted FBI training academy positions, and then the police chief's job at the Cape Girardeau department in the years ahead. In fact, briefly in the 1950s the duo were abruptly brought back to share power as *co*-Chiefs of Police in Cape Girardeau, an unprecedented state of city affairs. At that time, ex-boss Carl Lewis was also strangely brought back to act as Cape's fire chief once more. Coincidences?

The Mayor of Cape, thirty-year-old Wiley Hinkle Statler, if notified could well have arrived, perhaps with curious friends and/or family. It was such an unusual and monumental event, perhaps coming on the heels of a Saturday evening Catholic mass. "Hink" might have gone for the same reasons as Chief Barenkamp, maybe even taken his wife, if they had been together for a social occasion anyway. {Likely unrelated: the local newspaper recorded that just a few days after April 12th, the mayor's wife, Catherine, suddenly felt so ill she was hospitalized in Cape Girardeau.} It's of some consideration that W. H. Statler may have seemed like an obscure ex-mayor within four years' time and yet he was granted a position in the United States Navy; a Purple Heart in WWII; a private audience with the vice president of the United States in 1945; and some months after that, he was ushered in to see the very same famous Missouri politician after he rose in power to become the new American president. Also mere coincidences?

Cape Girardeau's City Attorney, R. B. Oliver, Jr., might well have arrived at the site since so much official police and fire business ended up within his caseloads on his desk. Teenaged R. B. Oliver III had just arrived in town for the holiday, according to the papers, and he was a friend and political pupil of a top Cape fireman, so the son might have tagged along, or at least heard the tale of what really happened too. The Cape County Coroner, Norval "Grub" Short, was likely notified and present too, perhaps along with the County Medical Examiner, Dr. J. H. Cochran. A physician (perhaps a few) and some nurses, since all were expecting casualties in need of immediate medical treatment, if not corpses in need of wrapping. Along these lines, a hospital vehicle - an ambulance? - with these medics, along with a coroner's vehicle with attendants. they might have been standing by at the site, or at least on their way in the dark countryside before matters were wrapped up. An unsubstantiated source of older hearsay (speaking in 2014) claimed that the alien bodies were scooped up that night and rushed to a Cape Girardeau hospital, where the nurses "turned white with fright" at the sight of them. It may be, however, that the nurses described in this allegation were only called *to* the crash scene, where they blanched at the shocking ET corpses in the grass.

Oddly enough, it is very possible that another eye-catching vehicle was parked near the farm crash site: a *hearse*. That's because, according to newspaper summaries, it was common practice in Cape (and other cities of that era) to utilize funeral home employees as emergency first responders in their company hearse, complete with oxygen mask and tank, cot, and first-aid kit in back. In the days before Emergency Medical Technicians, a hearse driver and attendant would have been given basic training in first-aid via personnel in the Cape fire department, according to a period newspaper report. As shaky and even comical as this sounds today, those whose business was dealing with dead bodies were often sent out to help at accident or medical emergency scenes, to help revive patients and rush them to the hospital. And if the victims didn't survive the trip. well, guess who handily had themselves a customer for their mortuary and funeral services?

Were there doctors and nurses present at the crash site? In a 2014 e-mail, a longtime Cape Girardeau resident recalled his sister told him the MO41 story from the start, in the days after the macabre event, via a friend that had the lowdown on the rural crash. What was recollected was that some nurses from a local hospital were able to view the alien bodies. This was quite memorable, it seems, as "they went white with fright" at the sight of the gray, weird cadavers they were confronted with. It shook them horribly, as it logically would anyone unprepared for the truth of what lay on the ground that night. And if there were white-uniformed female nursing aides at the site, then a trained physician was likely present as well, perhaps more than one, also imported from one or both of Cape Girardeau's hospitals. Perhaps some or all were affiliated with the local Red Cross, which helps out at various community emergencies.

There might also have been another unusual vehicle parked near the accident site: a mobile radio transmitting truck for a live broadcast at the scene. Cape Girardeau's main radio station, KFVS - later their TV station call letters to this day - utilized just such a wandering vehicle. The white, boxy "mobile unit" was manned by a driver/technician and an intrepid reporter, always in search of a good story for the area's airwaves, easily the main source of home entertainment and news for Americans in 1941. {Televisions numbered in the hundreds nationally and there was little to no broadcast content in the early 'forties.} Either the hot tip of "an airplane crash" or "a weird spaceship crash" would have brought out a reporter, eager for a live, on-the-air scoop. Evidently Minister Huffman did not specifically recall to his family such a KFVS reporter or his notable, well-marked vehicle, however, yet it should be remembered that it was quite dark at the site after the fire was doused and William was not there the entire time.

By now it was close to ten p.m. The spaceship the various onlookers openly inspected "was very shiny," William Huffman's grandchild recalled him saying much later. An amazingly "smooth metallic finish," when not crumpled and damaged, unlike any vehicle anyone had ever seen before. Grounded, rounded, and all were astounded. Huffman did not recall seeing or smelling any spilled fuel at the site, unless that was what was burning up in the flaming debris field. What fuel or energy source did this quite unconventional airship run on? How did it previously stay aloft and navigate without wings or propellers? No one had any answers.

Pastor Huffman must have gripped his Bible closer as he glanced briefly and blankly at the control center of the ship, at least at first, since it was attracting the most attention. Once again, the *bodies* were the most important point of interest for William. The victims. His reason for being there. Someone in the crowd supposedly asked the pastor to "provide blessings" to them, seeing as how he was not Catholic and was unlikely to have attempted any "last rites" on these strange dead critters. William probably thought at one point they looked like ugly demons straight from hell.

To the uninitiated, the whole scenario might well have seemed like the start of a possible otherworldly invasion, perhaps by "Mars men." It could well have appeared - at least at first - to be a real life version of the well-known

October 1938 radio broadcast of *"The War of the Worlds."*
That was the famous fictional production by Orson Welles
and his stock players, of a once-dull 1898 H. G. Welles
novel regarding an extraterrestrial event in the rural
countryside, a live play in '38 that caused a very small
number of well-meaning citizens to overreact with fear,
ignorance, and gullibility. Those mistaken, nervous folks -
mostly on the U.S. East Coast - called up police
headquarters and radio stations, friends and family around
the country, demanding to know more about "this Martian
invasion going on in New Jersey."

In reality, the American media in 1938 - and for decades to
come - greatly overstated the general population's emotions
and reactions to the alien-theme national radio program, as
some later careful studies and books showed. Only a tiny
fraction of the listening audience actually heard the faked
radio play, and only a few of *those* people actually (mildly)
panicked during and after its airing, mostly those in the
New York/New Jersey area. Yes, a few citizens there
anxiously armed themselves and some locked their families
in their barricaded homes, grittily ready to defend their
lives against possible weird-looking, hostile aliens,
perceived to be taking over and killing the human race. A
few more Americans took to the streets, scanning the
countryside and skies for the supposed extraterrestrials, but
that was about it. There was no genuine hysteria or panic,
organized militias or anxious deaths from the silly Welles
broadcast. It was the Depression-era *media* that managed
to produce a trumped-up but profitable delirium in the
aftermath, misreporting and exaggerating the small
percentage of negative emotional reactions around the
country, looking to sell more papers and gain extra listeners
for a few days. But the entire affair had legs. The
appearance of a panic in some sections of the nation helped
fuel the sad falsehood to this day that Americans would
react with mania and madness should they hear that ETs

have arrived on our soil. After two weeks of hype, the American media quickly moved on to other stories in the fall of '38. Still, the exploitive U.S. press - and proud Orson Welles himself - created a powerful lasting myth that millions of misinformed, fearful United States citizens everywhere had indeed panicked over aliens, for nothing. and perhaps would if they learned of such a tale in real life. The U.S. government and military took careful note of this.

It's quite possible that for many in America and around the globe, *"The War of the Worlds"* production was the first time anyone had given much if any genuine sustained thought that not only was humanity not alone in the universe, but that advanced alien astronauts could travel to earth in gravity-defying spaceships, study our cultures, and show technological superiority to mankind in almost every way. So the Orson Welles production served perhaps *one* useful purpose, even if it was a generally negative viewpoint, with the fictional armed "visitors" to American shores described as angry, blood-thirsty, and ruthless. Other comparisons at first are startling in similarity: both Cape Girardeau and *"The War of the Worlds"* featured an alien spaceship landing on a rural farm outside of an obscure small U.S. town. Both stories describe curious, then fearful citizens, the police, and the army, all driving out to the countryside in various vehicles, blocking roads as they parked, to nervously inspect the landed vehicle and alien beings in the back yard of an unnerved farmer's house. And both nighttime sagas played out, initially, on a weekend, after eight p.m., with repercussions lasting for weeks, months, and years afterwards. Near the beginning of his Depression Era "Mercury Theater" radio show of the British book, Orson Welles informed listeners solemnly, "We know now that in the early years of the twentieth century this world is being watched closely by intelligences greater than man, yet as mortal as his own." He couldn't have been more right, as the '41 Missouri crash soon proved.

In 1941, there was almost no television at all, it was still in its infancy; sets were owned by a mere few Americans with precious little content broadcasted. Thus, no creative programs about evidence for alien visitation were in existence, to help condition the mind to calmly accepting this notion, as is commonplace today. There were also of course no home computers for web-surfing, no cell phones or I-pads, no video games and apps, and almost no book or film references to space creatures, minus some cheesy *"Buck Rogers"* or *"Flash Gordon"* B movies, almost always as two-reel serials or quick, shallow, episodic fantasies with humans cast as otherworldly beings. Possibly a wildly imaginative comic book adventure like the brand new *"Superman"* series could briefly touch on the subject of extraterrestrial worlds and creatures, but even those were not widespread in the Depression. *The subject of aliens was just not mainstream in 1941.*

Reverend Huffman scanned around in the darkness but his eyes rested upon this one alien on his back in the young blades of grass before him, moving his chest up and down lightly as he wordlessly tried to inhale an atmosphere apparently quite unlike his customary own. "Granddad said that when he got to him his breathing was shallow," Charlette Mann stated firmly in an on-camera 2008 interview in her cozy Tyler, Texas, home. A fresh feeling - *pity* - undoubtedly rose in the gentle pastor's soul. This humanlike creature seemed sadly to be at the point of death, like his two compatriots. What could the preacher do to help? William tried to think, but if he were to act impulsively, not knowing the unconventional being's anatomy, well, what if he just made matters worse? "Mouth-to-mouth resuscitation" was not a common term or procedure in those days. The clergyman wasn't a trained healthcare professional. Matters of spirit, not flesh, were his specialty.

Between 10:00 and 10:15, it can be estimated, the preacher gazed around and listened in to the conversations of dazed, firemen and policemen; healthcare and coroner crews; eventual FBI and Army personnel; plus civilians and newsmen on the scene. This was history! *Before them was the first actual contact between humans and aliens in American history, possibly in all human existence.* The problem was, the only extraterrestrial left alive was not communicating, and barely moving, and no one knew what to do next. Perhaps some wiser minds knew to fan out and search a wider perimeter, to make sure there were no other creatures, thrown from the wreckage, still laying in the dark dead or alive (and wounded), or any second crashed craft nearby. It was all being recorded for posterity, William noted...

In the crowd, apparently *two* local gentlemen with sizeable professional cameras were getting some work done, William recalled to his family later that night. They seemed like newsmen, working an accident scene as they would normally, looking for the facts and recording the outdoor wreckage and the inside scoops. With trembling hands, no doubt, the duo snapped picture after picture, their bright flashes briefly illuminating the scene on an almost blinding scale at times. The hot, spent flash bulbs would eventually go bad and need to be discarded and replaced, as would slides of sheets of camera film, as was the style at the time with the big news cameras. The newspapermen felt they were capturing the wild scene for all the world to soon see. Perhaps they if could also sell some of the resulting images to a national magazine, well, what a sensational, lucrative story! Flashbulbs went off in the farm field, dollar signs went off in their heads.

Charlette Mann remembered seeing many times a small, stunning black and white photograph that was snapped at the nocturnal accident site, featuring one of the foreign-looking creatures in pretty clear detail, being propped up and posed by two plainclothes men (see Chapter Two). It was allegedly taken by one of the two cameramen, but while utilizing a different, "small personal camera," not the larger, more complex ones in use that night. The photographic results later were developed as "around three by four inches," give or take a little. At the crash scene, the winding, clicking, slide-pulling, and bulb-popping bigger cameras went off now and then, adding to the historic importance and surreal atmosphere, appearing almost like a big-time police crime scene.

One of these two more serious cameramen present is believed by some to have been thirty-six year old Garland D. Fronabarger, a respected local newspaper writer, reporter, and photographer. Angular Garland was a well-known man about town and a member of Cape Girardeau's First Baptist church, Kiwanis Club, and Lion's Club. The gruff newsman preferred to go by his initials, G. D. Fronabarger, but most knew him around town simply as "Frony." If the Cape police/fire personnel had called Pastor Huffman, it would make perfect sense for them to have called and tipped off *The Southeast Missourian*, who quickly sent their roving ace reporters to the site of what all felt would be a very serious story: an airplane down, victims in desperate shape, and heroic rescue efforts taking shape as precious minutes ticked off the clock. Frony was a very strong candidate to have been sent, even if called at his Cape home on South Park Street, after hours, where he often developed in a dark room some of his hobby photos taken from his small collection of cameras. He was an experienced, no-nonsense newsman who always lugged his trusty Graflex "Speed Graphic" camera with him and recorded pertinent facts and interviews by notes and photos. He knew his business well, plying the journalism trade for over a decade. Sometimes this entailed setting his rather heavy camera up on a tripod, and lugging extra slides of film and flashbulbs, often within a carrying case for all his journalism gear. Possibly other *Missourian* personnel also scurried to such a breaking major story of this magnitude, perhaps even including an editor, or the publisher/owner of the papers, the Naeter brothers, but to date that is unknown.

Speaking out in the aforementioned 2014 e-mail, one longtime Cape Girardeau resident revealed his sister recalled to him (and to other kin) many years ago that she learned Garland D. Fronabarger was indeed at the UFO crash site, numbly taking pictures of the strange ET victims and their damaged airship. Yet not a soul has come forward to say Garland Fronabarger even *hinted* to others at knowing about the alien affair. According to Garland's son, John, conversing in a telephone interview from his retirement home in Arizona in 2013, his father the longtime newspaperman didn't utter a peep about such a dramatic and traumatic encounter to his family. Yet John added tellingly: "Dad was secretive about some things, and talkative about other things." Thus there is a real chance Frony kept it all to himself, to spare his family any potential trouble from loose lips; he needed his job and his solid reputation badly. The springtime 2014 source stated that his sister related from the beginning that she had learned that G. D. Fronabarger was *there*, and that he was carefully sworn to secrecy by authorities after having his mind-boggling news-reporting evidence snatched away. Garland was also supposedly sternly informed *never* to speak of the incident if he knew what was good for him, something that conceivably could have been reinforced with follow-up phone calls over the years.

Further details of the special picture taken night and was later viewed by William Huffman's relatives - of a limp, presumably-dead alien - leads one to conclude it was likely snapped by G. D. Fronabarger (again, see Chapter Two). For instance, Ol' Frony owned different cameras and could well have kept a small reserve model in his pocket as described at the scene by Huffman. Taking pictures was nearly his whole life, for a living and for fun. One camera would have been for capturing news, the other was more likely just for his whims, as a "hobby camera." {Or perhaps was actually owned by a friend or relative and he had it with him.} The larger Graflex model he used in the Great Depression was considered the classic newsman's camera, and Garland loved his, using it almost daily for several decades, amassing an extensive photo collection even as technology changed and newer and improved models came out. The rather famous Speed Graphic utilized very large flashbulbs, which produced nearly *blinding* flashes of light when ignited. The old-style camera also required its users to get up close to subjects, with no telephoto lenses available, and methodically produced 4x5 and 5x7 inch exposures. It is possible Frony went out and purchased its smaller, lighter form, the "Miniature Speed Graphic," which came out in 1939, or a similar model. The "Mini" produced smaller prints, some of them 3x4 inches, the general size Huffman's kin recalled of the exposure souvenir from that fateful night. Or perhaps a modest Kodak "Brownie" type of camera, using small rolls of film, might have been in play, as Charlette suggested in interviews. This little device could have easily been placed in a large shirt or vest pocket, as the Huffman family story alleges, as clothes were often looser-fitting in those days.

Yes, the camera flashbulbs were popping here and there that night, intermittently lighting up the alien accident scene, despite the diminishing height of a few dying flames still barely left flickering in the surrounding field. Charlette Mann's account of Floy Huffman's memory of what her husband recounted was that an unidentified cameraman pulled the smaller personal camera "out of his shirt pocket" and took apparently *just one* black-and-white film's snapshot of the specific ET, posed by two men, as mentioned. A single quick image was Frony's specialty. Small cameras like this were not all that uncommon in 1941; for a man to carry *two* cameras on him was unusual and clearly indicates his strong passion for photography. Therefore, the small-camera source can once again be logically pinned on G. D. Fronabarger. "One Shot Frony" *loved* cameras and photogravure and sold his finest collected images to area residents from time to time, to augment his income, popularity, and photojournalistic reputation. This would be a real doozy, he no doubt believed.

At any rate, as William ministered to the lone still-breathing alien before him, two bold men nearby picked up and held a limp dead gray alien between them. They had plucked out one of the two prone corpses and stood him still to pose as a remarkable trio for a cameraman's personal camera, just as so many people do today with diminutive cell phones and digital cameras easily stored in one's pocket. It is possible it was felt then that posing *with* the alien also produced "scale," to help a viewer more easily determine later the size of the extraordinary creature being recorded. The camera bug nervously focused, aimed, and pressed the shutter button, all at the most astounding sight of his or anyone else's life. His flashbulb popped and lit up the scene again. People blinked and went back to what they were staring intently at.

Had any of the creatures spoken before they died? Had they made hand gestures, or relayed thought transference? Or left behind a notation, gesture, or a sign of some sort?

William did not say so, as he was simply not there in the immediate aftermath of the accident, but he did overhear a clue, according to granddaughter Charlette. These two dead non-terrestrials on the ground, side by side. "several said were dead on impact," Charlette recalled hearing from Floy, remembering William's words. They apparently showed no sign of life at all, and were evidently pulled away from the wreckage and the flames by first responders, maybe even by the farm owner, sometime before William and his police source arrived on the scene. But could they have been merely stunned, or unconscious? Who knew for certain with their alien anatomy?

Over the last evidently-living alien, compassionate William knelt in the damp grass and said some prayers - possibly to himself and/or some aloud - to the nonhuman victims and to God. Utilizing his Bible pages would have been difficult, as his hands were probably shaking and the light available to read by was constantly moving about all around him. Pastor Huffman would have had to largely wing it, his sad blessings and prayers. As bewildered William stared down between at the helpless victim, "the little man" (as he once referenced later) gasped for breathable air compatible with his specific respiratory system. He finally went limp, dead at last, William recalled mournfully later.

Huffman muttered more prayers for the deceased, pathetic beings. Were these "last rites"? That's the way Charlette Mann has repeatedly described what her grandfather was doing for the souls of the deceased alien beings present. Or was it just simple thoughts and recalled Bible verses that William silently utilized as he knelt, possibly shutting his eyes at times. The Catholic faith requires uttering "last rites" to those who people, anywhere on earth, who have just died and their souls are now beyond our world, perhaps in purgatory. To have their sins forgiven and hopefully absolved. Kindly William was a trained Southern Baptist preacher, but this was a simple matter, a process of praying to God for protection, enlightenment, and a swift trip to the afterlife for the spirits departed. Either way you describe the act of prayers for the dead, W. G. managed to accomplish the goal that was set before him, when the Cape police associate called him at home.

By this point, the fire had been all put out. The bodies declared dead. The prayers issued. The danger seemed over. The urgent "rescue operation" of helping airliner crash victims had become a rubbernecking curiosity site, nearly a pleasant social affair to enjoy, record, and discuss, but then ponder: *what do we do now with this mess?* Did it all simply belong to the farmer who owned the land, and was *his* problem to clean up? Could anyone who asked keep some debris as souvenirs? What if more of the creatures come back and want to retrieve everything?

Another important question must have arisen, and lingers to this day: *why did the alien airship come crashing down in the first place? What was the exact cause of this terrible wreck?* No one recalls there being any lightning-filled electrical storms in the area that night. No one recounts a collision with either a manmade or second extraterrestrial aircraft. And no one can say for certain if there was "pilot error" or "structural problems" or "mechanical difficulties" that accounted for the tragedy. No one seems to have recalled hearing a sonic boom that evening or seeing any sort of vapor trail in the sky. Southern Missouri's sloping hills and river-area soil was good for growing most any crop, but also for grazing by cattle. Could the aliens have been attempting to abduct or "mutilate" or more likely "harvest" a cow for its organs, as other UFO/ET cases seemed to indicate decades later, elsewhere in the country? Did they then run into simple mechanical or pilot error, or both? One can only speculate and imagine.

One thing is for certain, there was little up in the sky to tangle with in 1941. No orbiting satellites, no space program vehicles, no drones, no real helicopters (or crude "auto-gyros"), only a few big airliners, and precious few blimps or weather balloons were in use in America in that era. The reason why the alien trio's rounded craft plummeted and smashed into metallic pieces will likely always remain shrouded in mystery, along with precisely why that part of the North American continent was being flown over during the Great Depression in the first place. Was there something in or around southeastern Missouri that needed closer scrutiny by an alien crew in a small scout ship? Was it really a routine reconnaissance mission gone wrong? Or was it simply "passing by" planet Earth, in the upper atmosphere, when it developed a problem and sailed down to Show-Me State soil by mere chance, a random accident that had nothing to do with cultural observation at all? Or perhaps, wildly, it was shot at and hit by an American fighter plane or by hostile WWII fire somewhere in the skies above, and only came tumbling down to a crippled halt outside Cape Girardeau by mere chance. One can only speculate.

There's little doubt that in all the socializing, note-taking, photographing, and firefighting that night, the gathered men stopped to wipe the sweat from their brows and cool off. The day's forecast as shown in the afternoon paper was for a high of 84 degrees with only a light wind that likely died down at around 7:30 sunset and dusk. The sky was clear and thus with a high ceiling and great visibility. It would seem impossible for a mid-air *collision* to have occurred, as no report claimed a manmade airplane or balloon was down in the vicinity, nor a second alien crash site. Later on that night, with rattled William was safely back home, he told his spellbound wife and two sons a further startling fact: he stared carefully into the unusual, partially-damaged craft and *got himself quite an eyeful.* He had inside *details*, and was ready to impart them, flush with the exciting, tingling non-terrestrial memories.

The exterior of the busted ET ship had a very shiny and smooth metallic finish, shaped "like a saucer," as Mrs. Mann phrased it. The interior command module or flight deck was the same general color, perhaps illuminated by the flashlights and flashbulbs blazing nearby. No great odors or sounds were reported in this internal central, likely stabilized cabin, so it's likely none were present. With the aid of these lights, William could make out the interior crew compartment and its contents. "There were some little metal seats, children's sized, before an instrument panel," the dazzled evangelist/fund-raiser recalled. "And writing on a metal band all around the inside of it, with weird symbols," Huffman recounted to his shocked, spellbound kin. The odd "writing" was like Egyptian hieroglyphics, with inexplicable little pictograms or symbols he simply did not recognize at all. There were clearly small gauges and dials inside too, and still other indescribable aspects that astounded and confounded William wasn't quite sure about. The parson did not report seeing any beams or seams, rivets or screws, nuts or bolts, or signs of welding or soldered joints. It was all so smooth, so unlike anything he had ever seen before, but a considerable section was ripped apart, with its silvery shards sprinkled around and sparkling in the farm grass. What exactly caused this explosive, messy destruction was not clear, other than perhaps the hard impact of the collision releasing flammable chemical reactions from the "power plant" within the engine, mixing with the oxygen-filled atmosphere of the earth. Whatever caused the destruction of this part of the advanced disc-shaped craft rendered it inoperable and immovable, from the aliens' point of view. Soon it *would* be moved, but not in ways the big-headed extraterrestrials were counting on. The evangelist had peered into the largest section of the crashed disc. Suddenly more lights appeared in the night sky and pierced the crash scene. Metallic vehicles were moving in. People froze in fear and confusion. Engines and noises could be heard, getting closer. Invaders had indeed arrived at the crash site...

According to the Huffman narrative, uniformed United States Army members pulled up to the scene in trucks and automobiles. The ponderous fun and games were over. Solemn soldiers - likely with guns - jumped out of their hastily parked vehicles and fanned out. Reverend Huffman stepped back from the unconventional craft and looked around, probably fairly startled. The strangers just barged in and took over, mostly young men in yellowy-tan garb, ready to implement their will on every player involved. This was a matter of "national security," the ranking officer in charge loudly informed those gathered around the crashed disc and bodies. *"This did not happen*, do you understand? You are to turn over all evidence from this incident, and never speak about it again. To anyone, any time." Many evidently heeded this warning. For one brief interlude later that night, Preacher Huffman didn't.

"Give us all of that debris and step away from those bodies," was more than likely the next stern command to the crowd. "Our orders are to take *all* physical evidence and reports from this area, and to have your patriotic assurances you'll remain silent for the good of our country," was probably how it was phrased, or very close to that effect. By now it was likely around 10:15 to 10:30 p.m., possibly just after that.

It was obvious the army/aviation boys weren't messing around. They had encircled the site quickly and moved in, closing it off to outsiders, shouting instructions, showing they were well-informed as to the ET nature of the incident and meant business. It was almost as frightening a memory for onlookers as those of the spooky alien bodies. All of this takeover was possible, we recall, as "uniformed men from the military" were *already* present at the site when William arrived; they were calm and had likely called for reinforcements before the Baptist preacher/fund-raiser was even notified by phone to go. Orders from a high level of command had evidently been handed down and impressed upon the entire arriving U.S. Army personnel. The almost-festive mood and excited smiles for all locals melted away fast, everything grinding to an abrupt halt. Soon it must have seemed nearly like they were under mass arrest, and for just being innocent bystanders at a vehicular crash.

The rattled reverend stated he was taken over to one side of the farm compound and commanded to keep his mouth shut - forever. William could only logically assume later to his family that the others at the site were also told the same thing, herded off "in small groups" to the smoky, dark edges by the uniformed military reps. That W. G. Huffman did not actually see or hear much of this process strongly indicates that he - and perhaps his driver, the dutiful cop - left the scene not long after his secrecy oath and warnings. The rescue had turned into a recovery operation. It was time to for all others to leave, the party was over. The Cape police officers - even in casual attire - likely would *not* have enjoyed being bossed around by younger, uniformed Army personnel. William and his police associate pal - Justice Cobb? - scooted back to their parked car by the roadside, the one-fourth-of-a-mile walk that was now more difficult, thanks to all the other visitors and the new military vehicles in the vicinity. Likely even pulling the automobile onto the dark road was now problematic, with others trying to leave and military men now swarming the scene, patrolling the perimeter, eyeing all suspiciously.

Before nearly all departed the scene that night, likely between 10:30 to 11:00, it seems each individual present was ordered to drop to the ground or hand over *all* physical evidence of the crash. Apparently some were likely searched. It seems pretty assured they all had the issue of complete secrecy verbally reinforced, perhaps even by a sworn oath, although one man - Garland Fronabarger? - evidently took his oath with crossed fingers, his small personal camera still buried in his pocket, safe from discovery. It was for the good of the country for all present to clam up, permanently, that was the Army's message again, but a little souvenir picture on a roll of film. well what the soldiers didn't know wouldn't hurt 'em.

Clearly the military thinking was: we don't want to alarm any American citizens *anywhere* with this shocking news; they didn't need another "*War of the Worlds*" panic on their hands. And they didn't cherish any possible propaganda coups for any foreign powers, regarding garish media publicity for a crazy-sounding incident that would make America's leadership look bad. And in 1941, United States citizens generally did everything their representatives in the government and military ordered them to. If told to shut up for patriotic reasons, nearly everyone informed as such *would*. It is quite possible that the armed forces personnel present made it clear to everyone gathered that there would be unpleasant repercussions if the eyewitnesses did not keep the event hushed up. A veiled threat, or perhaps even an open, obvious one, should an indignant eyewitness have responded negatively, or even with rebellious sass. Almost no one wanted to rock the boat or make big trouble for oneself and one's family. No one really wanted to be thought of as a kook or a prevaricator, or rabble-rouser by blurting out the tale later. Keeping quiet for life was easily the most logical, sensible move for every person present. *It was best just to forget it.*

Charlette Mann mentioned how as she understood matters from her grandmother, people from Cape Girardeau who knew too much were genuinely scared into silence in the aftermath. Some citizens apparently began to change in personality and demeanor, a few even in a positive way, including Reverend Huffman himself. Supposedly, he became more tolerant and open-minded, but deep down still rattled, according to Floy. The shaken minister who witnessed the celestial world, in effect, went on to become "more accepting of the unexplained," and less closed-off to the paranormal. But for him to say - or for his wife - that other visitors to the scene altered their moods and behavior in the wake of the tragedy and its recovery op indicates that the Southern Baptist couple had contact with these local citizens around town, or heard stories about the unusual change in their conduct as life in Cape Girardeau naturally went on that spring and summer of '41. This meant the clergyman recognized names and positions of fellow witnesses at the crash scene, and could have imparted still *more* details to his family, but opted to use discretion instead.

According to the remarkable Huffman family memories, other Cape citizens became *less* open to any sort of discussion of anything metaphysical. Mention of the crash sent some scurrying away. Perhaps they had good reason to shut down interest in hearing or talking about anything remotely linked to life beyond earth. The consequences scared them greatly. Who could blame them? Why, just by speaking seriously about "alien creatures" and smashed "spaceship debris" to the uninformed average American in those days... such storytellers would have been laughed at and scorned. *Alien people*, from other worlds? Flying *round* airships? There were no such things! Or, only found in wildly fictional "dime store magazines," action-adventure stories for kids, or "pulp fiction" as it was better known as then (and to this day). Newspaper comic strip adventures and a few crude Hollywood B movies on "Martians". that was silly fantasy. As mentioned, far fewer people in those days believed in intelligent races beyond mankind, or planets with intelligent civilizations beyond Earth, or even discussed such oddball topics. No one wanted to be thought of as unstable, delusional, or a liar.

Beyond this, Cape Girardeau hosted an armory. Citizens who joined the Army National Guard were allowed to work and live in the area and tend to their military duties on specified weekends and retreats. Thus, U.S. Army personnel were active in the community, going about their business in everyday civvies, and they had eyes and ears open to those who might have mentioned the crash. It would have become difficult to know who to trust, and to whisper stories to, even behind closed doors. Who might rat on you later? It was very much in one's best interest to keep lips zipped and not even jot down notes on paper, or in a diary, for future relatives to get into trouble by possessing.

As mentioned, there was one American citizen who would undoubtedly need to be called and notified of the stupefying Cape Girardeau situation: President and military Commander-in-Chief Franklin D. Roosevelt, who was kept up unusually late in the White House on the 12th, into the 13th, according to records. FDR's famous wife, Eleanor, suddenly and mysteriously had her flight canceled in Boston, and had to hope a train back to Washington, overnight. Franklin's top military aide was the trusted and respected Army Chief of Staff, General George Marshall, and his travel records also show a possible break on this night and into the next day. For the record, he seems to have gone "missing" until Monday morning, when he showed up at his D.C. office. GCM was likely alerted that night and thus passed the shocking Missouri information up the chain of command, to FDR and perhaps War Secretary Henry Stimson. Command decisions were needed, for *who knew just what was going on?* Had an alien invasion begun, and the farm field outside Cape was a telling accident that inadvertently tipped off the American armed forces that more was to be expected? Was the crash in reality part of a first wave, or reconnaissance invasion probe sent out? Was it time to panic? It was anyone's guess. But one thing was sure: *the crash was now a matter of national defense.* The U.S. Army was now naturally in complete command of this emergency Cape Girardeau situation, so full of unknowables. Urgent calls - some plausibly made on military radio equipment - were undertaken in as private and controlled settings as possible, perhaps even from an emptied farmhouse or a public pay phone that night.

It seems logical to believe that the mind-set of the military that night was along the lines of: "For the good of the nation, and maybe the entire world, we've got to contain this, investigate this, and cover this up, at least until we get some definite answers. *It might be the start of an interplanetary war.*" Who could blame war veterans in charge for thinking such a thing at that uncertain moment?

Back in Cape Girardeau, the somber soldiers ran off the final sworn, scared crash witnesses. The farm was largely cleared, minus the property owners. Then the Army set about to scoop up and haul off with every piece of disc wreckage they could find that night. This process undoubtedly extended into the light of the next day, or even later. And of course they had to deal with the three odd casualties. They were bizarre trophies, but a rotting corpse odor might well have been setting in as the warm hours passed. The rather grotesque-looking interstellar beings needed something more than a mere blanket or box to be stored in. Apparently also confiscated that night were reporter's notes and the photographers' camera film. Police report notations and fire report observations were also snatched up, perhaps any FBI notes were also absconded with, if Arlin Jones arrived. The stern Army attitude supposedly was: "*This did not happen. It is not to be recorded in any way, not even in response logs.*" No news article was to appear anywhere, no record was to be left of it every having happened, that was made clear. An imposing U.S. Army member probably went later to an editor or the publisher of *The Southeast Missourian* newspaper to ensure that no printed stories or even subtle hints would appear at any time. It may have seemed to any newspaper person like the story of the millennium, an electrifying bolt of lucrative news to capitalize on for weeks or even *years* to come, but it was very effectively quashed. Government censorship over the rights of "Freedom of the Press" was substantial then; reporters were not even allowed to reveal in the media of the day that their own president was largely confined to a wheelchair and could barely stand up with help for a few minutes at a time.

The military personnel present at the scene likely also talked, at some point, to neighbors and relatives to the farm crowd. Or even any curious passers-by who overheard too much. The soldiers were thorough and intimidating, precise and stern, and got the job done. No one dared to cross them. But who precisely *were* these soldiers, and how could they have arrived within just a few hours of the crash-landing? There are two main candidates.

U.S. Army representatives used to run a recruiting station in Cape Girardeau, but that was closed, perhaps temporarily at least, in mid-March, so stated the local papers. For many decades, as mentioned, the city had been the site of an Army National Guard training center, nowadays a hilltop armory in town where recruits are trained and armed for action, in peacetime and in war. In fact, the Cape armory was where a bookish young enlistee named Harry S Truman trained in the summer of 1906, living in Cape Girardeau thirty-five long years before the marvelous

Missouri mystery, a decade before he was shipped off to Europe to fight as a captain in World War I.

Industrious Mr. Truman in April of 1941 was a highly respected U.S. senator, probably in bed by 10:30 p.m. Eastern Time within his apartment at 4700 Connecticut Avenue in Washington D.C. Ironically, Truman was a friend to none other than Cape newsman Garland Fronabarger. Hardworking Harry was at this time was heading up his own military oversight senate committee, trying weed out waste and fraud in contracts and services by and for U.S. armed forces. Harry's first witnesses sat down for a grilling on Tuesday, April 15th, and included high-ranking U.S. Army personnel, the Secretary of War, and General George C. Marshall. In fact, to others and in his memoirs, Truman mentioned how he grew most impressed that April day with Marshall, in conversations before, during, after his sworn testimony on Capitol Hill. Marshall also took time to attend a top level meeting with President Roosevelt at the White House on this Tuesday afternoon, surrounded by Secretary of State Cordell Hull and Marshall's boss, Secretary of War Henry Stimson, along with presidential aide Harry L. Hopkins and two other officials. {These names become more fascinating and revealing as MO41 unfolds.}

The next day, General Marshall was back, meeting with top military brass and FDR (plus Hopkins) in the Oval Office, where logs noted the unusual declaration *"off record."* Meaning that the subject matter of the closed-door confab was so explosive and classified it was to be considered by all participants as "off the record," with no notes or dictation allowed. Just four days, apparently, after the wild Cape Girardeau affair, time enough for Army photos to be developed and briefings typed up. This huddle took place just a few hours after the president met at 1600 Pennsylvania Avenue with his personal physician, the logs reveal, along with his top research scientist, Dr. Vannevar Bush, another important name to keep in mind. But to be certain, the Missouri farmer-turned-senator Harry S Truman was *not* on the Oval Office guest list during this April week.

As Truman knew firsthand, the United States National Guard is a military reserve that does not rest, in effect; they are always on call for emergencies, to be ready to deploy at short notice in order to help out not just for military action, but for domestic national disasters, such as at tornado and earthquake sites. Even Harry was still an ex-Army officer "on call," of sorts. According to recent Spring '41 news accounts, while he was preparing to interview military leaders at his upcoming congressional defense hearings, Senator Truman was still considered a U.S. Army "Colonel of Field Artillery Reserve," as was his own personal secretary at the U.S. Capitol Building and Senate Office Building. Thanks in part to his 1906 Cape Girardeau armory training, Harry was an Army artillery captain in Europe during WWI. As of April '41, he was often either at the defense budget hearings or at his senate office desk "at eight a.m., sometimes working until late at night," the nationally syndicated news story described. However, for the past few days, and for over a week to come, congress was on a Passover/Easter holiday break. Almost every other U.S. congressman had returned to his home district. Hard-working Harry stayed in D.C. and kept prepping for his complex hearings after everyone else went home. During this time he was hit by extremely painful gallstone attacks, one evidently on the night of the 12th that he and wife Bess thought at first was a heart attack. His Commander-in-Chief, and fellow Democrat, the often-ill President Roosevelt, was also up late on Saturday the 12th, until after midnight, as mentioned.

At first, telephone calls (through operators) and radio messages to superiors - but not likely to Senator Truman - could have been made that Saturday night by any member of the impromptu gathering at the crash scene, or by police or fire department members back at the brick, three story station the two departments shared on the corner of Frederick and Independence. Some of the CGPD and CGFD may well have been national guard members who were aware of the true nature of the otherworldly encounter and knew military-trained men in uniform might well be needed, if not right away, then in the aftermath, if more aliens were to come in and try to retrieve the remains and wreckage of their fallen compatriots. But the National Guard in general is not deployed by a civilian order or even by a mere local officer involved with its formation in his community. That special order would normally have to come down from the head of the state, Forrest C. Donnell, the new governor of Missouri, residing at the executive mansion in Jefferson City, hundreds of miles to the north and west of Cape Girardeau. No records exist at present to prove Governor Donnell took action of any sort that night, but if he had, it is possible such logs were later scrubbed, or never properly recorded in the first place, once again owing to "national security."

Neither is it logical for someone to have placed a call to the United States congressional representative from the area, Orville Zimmerman, a friend of Truman's who was visiting his own family down in Kennett, Missouri, that Easter weekend, many miles to the south of the bizarre accident scene. Congressman Zimmerman was very much like Harry, a fellow New Dealer who worked in the U.S. Capitol Building after having served in the Army a quarter century earlier, and was a friend to newsman Garland Fronabarger. Whether Orville ever found out the truth is questionable, but. who knows? He was recorded by Cape's newspapers as attending some meetings in town on

Monday the 14th, but whether he visited the crash event two days beforehand is a mystery.

Another factor to keep in mind is that the army did not come raggedly rolling in that fateful night and only casually look over the scene. "They surrounded the site, at first," William remembered to his two riveted sons and wife, "and moved in quickly." The good reverend did not mention recognizing any of the well-organized soldiers who performed this duty. Thus, it is more likely a trained and methodical unit from just outside the city was deployed, one that had little or nothing to do with Cape's citizen-soldiers. The military unit on the ground was obviously given specific orders to surround and hem in the crash site and its onlookers, and then scoop up all evidence to whisk away, no matter how small. Those directives had to come from someone fairly high up in the army's chain of command, briefed on the situation and decisive in his response. Whomever it was specifically that had arrived upon direct orders, their first act was to "set up a perimeter," but whether they toted guns at the ready is not known. There may have actually been only about ten or so soldiers at minimum utilized to accomplish this search-and-recover mission, many more at maximum.

The second possible source for the organized military interference was much more likely: Sikeston-based army cadets and Air Corps officers from the Missouri Institute of Aeronautics, 309th Detachment, where as previously mentioned the Cape County sheriff had a brother working in their flight training office. Ben B. Schade, the Parks Air College employee who acted as the MIA's acquisitions manager, might have been called directly - or possibly via a base dispatcher - and quickly passed the staggering crash information on up the chain of command. This MIA base was situated perhaps only twenty-five or so miles south of the crash site, filled with trained soldiers, pilots, and a small band of military personnel who knew how to gather equipment and hurry to various locales on short notice. That's what military men train for and at times practice.

According to records and news articles at the time, Captain Charles Berton Root, age 31, was in charge of the Sikeston MIA flight school, where "hundreds of young men trained as Army Air Corps pilots." Just like Sheriff Schade, Captain Root had recently become engaged to a local girl, and was planning to wed a Sikeston school teacher on May 31st, according to the society columns of the local papers. Originally from South Dakota, C. B. Root had eight years of service in the Army Air Corps under his belt, stationed most previously *in Washington D.C.* If anyone from the United States Army was in charge of the air crash scene outside Cape Girardeau, it was most likely Captain Root, the man who daily supervised the military flight training operations in Sikeston. He may have called that night to his old D.C.-based superiors - likely very high ranking and well-connected officials - to ask the veteran commanding officers what they, or *he*, should do about the crash materials near Cape. The upper echelons of government and military always seem to repeat the same mantra: cover it up, whatever it is, in the name of "national security." This term was allegedly used by one or more Army men at the accident locale. They said and did what they had to in order to gain unquestioned control of the bizarre situation.

James Lightle was allegedly a key PAC/MIA player in the Sikeston training unit as he was an intelligence officer probably taking notes - or taking *away* notes - at any crash recovery operation, according to author Ryan Woods' research. James handled MIA logistics and transportation of materials in his normal daily duties as well, probably in conjunction with office-based Ben. Another important figure was undoubtedly imported from the MIA airbase: the flight surgeon, plus perhaps his assistant and/or a medical corpsman.   The unit would need their professional assessment of the alien cadavers, and of anyone who was subject to physical illness or psychological disorders in reaction to them. It was most probable that anyone who called the Sikeston MIA training center specifically mentioned there were weird dead bodies of unfortunate victims present at the accident site, creatures probably not of this earth.

The Sikeston-based air corps training detachment had only been in operation since the previous September, with the latest class of over one hundred cadets starting their training in late March of '41. The flight instructors - almost forty of them - were experienced in getting the young pilots ready for wartime flight operations via rudimentary air exercises.   The MIA primary trainees would do their more complicated flight maneuvers for combat readiness at other airbases around the country, especially after World War II commenced later that year. If the southeastern Missouri spaceship incident occurred on a Saturday night, however, many of the contingent might well have been off the base, enjoying down time at nightspots in the area, or have even been away on a holiday weekend pass, out of the region entirely.

The very modest Cape grass airfield - about to be fixed up and dubbed "Harris Army Airfield," later "Cape Girardeau Regional Airport" - was not too terribly far from the crash site, evidently. Unfortunately it had no landing lights and thus did not operate at night.   In fact, neither did the Sikeston airfield, which was largely a civilian operated. There *was* a nighttime airfield in Advance, Missouri, not too terribly far away, but it was no military airbase and was infrequently used at night mostly for cargo plane stopovers and refuelings, featuring rather crude sod runways. Therefore the controlling troop at the Missouri crash scene would have settled into available military transportation at their base, then been driven to the site by personnel who knew the area's roads and routes. They sure didn't fly to nearby the Cape airfield that night, although they might have the following day, in possibly circling the scene from above to fully assess the unique and perhaps genuine unnerving situation. Was our planet about to be attacked and invaded by a strange species? Were there other flying discs in the air, or on the ground in the Cape area, or elsewhere?

Whomever managed to arrive at the "Huffman site" took over, around sometime between nine thirty to ten thirty o'clock, it can be reasonably estimated. Perhaps ten, rounded off. Since Pastor Huffman did not mention any "flyovers" that night above the obscure, off-road, rural crash site, we can safely assume that no military personnel buzzed the flame-doused locale from above and then landed their plane in a nearby unlit farm field for fast action on foot. No, they were said by Reverend Huffman to have come in by foot, which means first via trucks and cars parked on the roadside. Then obviously the soldiers hopped out to spread out in the smoky darkness, listening to the voices and noting the movement of the crowd assembled beyond the parked cars and slowly budding trees, the fires now all extinguished.

At any rate, the Army soon clamped down on all discussion of the incident, including phone calls. They wanted - and got - complete control of the spooky crash and silence from its participants. No one evidently had the nerve to stand up for themselves and assert their civil liberties and right to Freedom of Speech. The young soldiers had guns and grit, and now they had the goods. The only thing the farmers and local first responders ended up with was the smoky smell on their sweaty clothes and sharp warnings ringing in their heads.

One noteworthy vehicle that was likely on the farm crash scene that night, nearly from start to finish, was the large new long-laddered Dodge fire truck the mayor had recently acquired for the department. Perhaps two, in hindsight, if other communities responded. Cape's fire engine - mildly damaged earlier in the day - was likely not just parked on the roadside, but pulled up to the crash as close as reasonably possible. The Cape Girardeau Fire Department responded with speed and dedication, under the direction of Captain Carl James Lewis, in charge for only the past four months. For an expected gruesome, chaotic "airplane crash" more than just the initial pairing of firemen, but more out of the force of eleven men, probably showed up. Plus another small town's fire department might have arrived and participated, such as some Chaffee or Benton, Missouri, personnel. Yet once again there is no definite recorded proof of this happening. No fire department's logs have been produced that reinforce the farm incident, but as according to Huffman's memory - via Floy and Charlette - the military ordered all present to complete silence. They were obviously not allowed to record in any way anything about the shocking incident.

The Cape Girardeau Fire Department call sheet does not show any April 12th response, not even a noontime "grass fire" firemen battled as described in a newspaper article, when the new fire engine was damaged in a traffic incident. An April 14th Cape newspaper article airs a similar officially unreported example. At 11:45 p.m. Saturday night, the news story declared, the Cape fire force finally showed up at a garage fire downtown and by the time they arrived the building had burned down and collapsed, damaging two vehicles inside. The location given would have been a very short trip for the fire personnel from their station house. unless the firemen with the hook-and-ladder truck were out on a call in the countryside south of the city. Meanwhile the existing CGFD's evidently-scrubbed or rewritten record of "Fires and Alarms" is suspiciously empty from April 6th to the 18th, while the rest of the month appears fairly busy. Thus it is very likely firefighters' report notes at the scene were taken, and those that weren't were later stolen and/or whitewashed. Logs might well have been rewritten at headquarters under stern, obeyed orders.

One of the well-trained firemen under Captain Lewis' command that night might have been Walter Reynolds, bravely fighting the flames around the crashed disc. That's because his grandson, Phillip Reynolds, Jr., claimed it so. But was it the truth? In late 2012, Phillip apparently spoke to paranormal investigator Stan Hernandez, who attended a UFO abductee gathering, where he alleges he met young Phillip and recorded his supportive 1941 ET crash story. In a supposedly brief, rushed interview, Phillip told Stan that grandpa Walter "was dying of stage four cancer" some time ago and told his son - Phil's dad, Phillip, Sr. - just what took place in that farmer's field way back in 1941. It was just as Charlette Mann said, Phillip, Jr., explained. Or rather, just what W.G. Huffman claimed; the pastor's story is true. At first, Walter stated, the Cape Girardeau Fire Department was informed there was an airplane accident outside of town. They received the address and sped to the site with an experienced, dedicated crew, including Mr. Reynolds. All were likely suited up, gear packed, and fully expecting a devastated airliner, blazing with jet fuel in the dark countryside, possibly with many victims on the ground in need of care.

Once at the impact site, the Cape fire personnel were predictably stunned at what they saw. The odd crashed disc and shiny metallic debris, with a series of small fires and local citizens, were all scattered about a field in the warm night air. This circular airship. this was no airplane with broken wings or uniformed pilots, nor passengers and crew waiting for urgent assistance! Walter Reynolds and his crew were instructed to put out the flames here and there just as they had been trained, and did so. Reynolds allegedly stated as he lay dying that he had decided during the 1941 emergency response to sneak closer to the damaged aerial craft, to peer in and see if there were any more signs of fire, or if there were any victims within the downed vehicle. Typical experienced firefighting duties. At that time Walter supposedly stared numbly at the three dead alien bodies near the ship, on the ground. The trio of little nonhuman corpses were just as the Cape Baptist pastor/fund-raiser described, and much the same as later "alien abduction" reports recounted, in the 1980s and '90s. Gray, spindly, with big round heads and black eyes. Lifeless and yet bloodless, unlike anything he'd ever seen before.

At some point, as mentioned the military arrived and began taking over the scene, which eventually left the dazed fire crew a bit on the sidelines, the small blazes now out and the strange crash victims beyond their reach. Walter Reynolds supposedly stood and watched as the dutiful military men hauled the disc away. He was told, "What you see here stays here," as all likely were when warned by the somber soldiers. As the uniformed military crew - presumably the U.S. Army - took control of the situation, fireman Reynolds decided to mildly defy orders by impulsively picking up a piece of metal shrapnel and slipping it into his pocket, as a tangible souvenir. This proved to be a rather dreadful mistake, as we shall see.

It's all a fascinating claim, very supportive of Charlette Huffman Mann's account, and very possibly completely true. but there are some problems with it.

Could young Phillip Reynolds simply have been wanting attention and made up his confirming tale? Or did Stan write up the story wrong, or exaggerate matters in his online article, where other UFO claims of his often appear dubious at best? At present there seems to be no hard evidence that a "Walter Reynolds" was on the Cape firefighting force in 1941, mostly because CGFD headquarters long ago ditched records from the 'forties. But also, Walter is not listed in the Cape Girardeau city telephone directory for 1942 (covering the previous year's "names and numbers"). No one has come forward with support for Phillip Reynolds, Jr.'s claim, including Phillip, Jr., himself, yet Stan alleges he found proof that Walter Reynolds was indeed on the firefighting force. then does not reveal what that "proof" is.

In the years after this 2012 report, there's been no apparent "cashing in" on the Phillip Reynolds story, no follow-up of any kind, evidently. Not by Phil Reynolds or even Stan Hernandez. Nothing exploitive or even supportive at all seems to have come of this lead, which in a sense backs up its validity. Why make up such a claim for money and not produce a book, or a video, or an online site ready to milk it to any degree? What was there to gain for either Reynolds or Hernandez?

The Stan Hernandez interview with Walter's grandson did add an interesting detail not heard previously in Reverend William Huffman's confession. Fireman Reynolds supposedly swiped this one small piece of spaceship shrapnel, found lying on the ground. In doing so, he tried to pocket it discreetly but was seen by military personnel. Walter was quickly reprimanded. He pulled out the metal and held it up; quickly his otherworldly evidence was snatched away by a soldier. Walter was then booted from the site. Perhaps fearful that a bitter, vengeful Walter Reynolds might talk, someone in power - Army intel agents? The FBI? Cape police? - allegedly bugged the fireman's telephone and began observing him as he went about his business in the small river town. Or at least, that was how an aging Walter Reynolds allegedly recalled it, looking back as the grim reaper closed in. He evidently felt back in '41 that he was being watched post-crash for any sign of rebelliousness and talkativeness about the UFO incident, possibly indicating that Walter became heated at being caught, reprimanded, and kicked off the accident site, and that those in power feared some sort of revenge on his part. Phillip Reynolds did not say that his grandfather gave the government or the military any genuine reason to shadow him or to worry further about his reaction to the eerie crash experience. Walter Reynolds was "a man of honor," incapable of making up wild allegations, his proud grandson assured researcher Hernandez, but the rest of the world can only speculate.

Could Phillip Reynolds in 2012 have provided us some critical affirmation of the stunning, pivotal ET crash story so desperately needed over the years? Or was he simply jumping on the bandwagon, glomming onto a somewhat popular online UFO allegation for his own fame profit someday? Either scenario is quite possible. Nothing about the Reynolds claim is demonstrably false. But it is possible Phillip just wanted attention and simply made up his story about grandfather, then made a quick exit because he wanted no further investigation or follow-up that might show his story to be a lie. The nature of the old man's cancer, a deathbed confession, and later revelations by a grandchild to a UFO investigator. it all sounds spookily and/or suspiciously like the Charlette Mann retelling of her grandparents' tale. Phillip has no hard proof, because his father had no hard proof, because Walter Reynolds was allegedly allowed to *keep* no hard proof, making matters very frustrating for all concerned. Stan Hernandez assured his online site's readers he would dig up more info on the exciting Reynolds allegement, but has as yet to come forward with any advancement of the story, on his website or via attempts to contact him.

Phillip Reynolds, Jr., calmly posed for a photograph to accompany the 2012 Hernandez article, before supposedly rushing off at the UFO conference, providing no backup details to his interviewer, or apparently anyone else since. It all remains a tantalizing but unresolved angle of our mind-numbing saga. But if the Reynolds story is somehow, someday proven factual, Reverend William Huffman - through his descendants - winds up looking more and more rock-solid trustworthy and dependable in his revelations.

Another Cape Girardeau fireman became truly fed up and turned in his resignation abruptly, just days after the crash, according to the city's newspaper. Small articles in *The Southeast Missourian* for two straight days mentioned Mr. Quintin Williams leaving the fire-fighting force and handing in his walking papers to a city clerk, but never discussed the reason for his seemingly abrupt departure. It seems possible and plausible Quintin was upset about the handling of the extraterrestrial incident and how responding officers were being warned into silence. Or perhaps by the shadowing going on in the aftermath? Williams might have been very insulted by the treatment of Reynolds, and perhaps Quintin was *also* being observed and phone-tapped. Maybe he had seen and heard and known too much. Or *done* too much. Sheer speculation, yes, but based on what we know now, it would seem as if Reynolds and Williams could have been the first responders on the fender-damaged fire truck at the UFO crash scene. Quintin Williams was the political friend of the Cape Girardeau City Attorney's son, and a devoted organizer of the Young Democrats club. Did some sort of political pressure play a role in Quintin's departure? We'll likely never know.

A perturbed Preacher Huffman was quite visibly and understandably moved as he sat back on his living room sofa, emotionally spent. "Grandmother said the look on his face was different," Charlette Mann remembered. As he wrapped up his spine-tingling tale, spent William was a little less ashen than when he first arrived home, a little calmer, but he was changed for good. At perhaps around 10:15 to 10:45 p.m. he had been driven home, he said, probably by his police contact, taking his Bible, his memories, and his somber oath with him. A stop at police/fire headquarters and/or at his ride's house - just across the street? - might have taken place in the late hours, say by 11:15 to 11:30. But finally, William was back at his North Main Street home, up the hill from the Red Star Tabernacle. As he reclined on his couch around 11:30 to midnight, surrounded by loved ones, familiar faces, the shocked pastor in the smoky clothing was probably bursting with the news. He pondered his vivid memories of the flames, the people, the disc, the shrapnel, and those creepy cadavers. What an incredible story! He had to have been torn in two: talk about everything that happened. or keep his mouth shut as he was just sworn and/or warned to do. William chose to talk, and lighten his troubled soul. Americans today can be grateful he did.

Speaking aloud in private what he experienced was therapeutic, good for Reverend Huffman's psyche. Having to break that oath of secrecy was likely difficult for a man of God, but the urge to purge - to his curious, worried, trusted family circle - was simply too great to resist. They needed to know where he had been, and why his expression was so pale and disturbed. And he needed to unburden himself. The "midnight confession" made all the sense in the world.

Probably around the time his clock struck twelve on April the 12th, and Easter Sunday the 13th arrived officially, emotionally drained William Guy Huffman concluded his tale as his dumbfounded kin finally released the edge of their seats in the quiet house. The wife and two sons, Guy and Wayne, were of course understandably stunned and spellbound by his incredible first-hand experience. Knowing the grandfather's solid reputation, they had no doubt he was telling them the absolute truth as it knew it and experienced it. They believed him. but it had to have taken a while to fully sink in. Weary William likely stood and trudged upstairs to bed. It was finished, in his mind. He apparently never spoke of the stunning event again. His two sons did, maybe his daughter-in-law, but that was nearly it.

For Floy and her sons, it was a relief to have William back, and not traumatized at least by the sight of human corpses, burned or dismembered in what they thought was going to be a dreadful airplane accident. But the tale the patriarch told! What a shock now to *their* systems! How could *they* now possible settle down and go to sleep for the night, like nothing had happened?

The Huffman brood naturally held onto the 1941 blockbuster tale long after patriarch W. G. Huffman, Sr., passed away in 1959. How could they possibly forget it? The forty-year Southern Baptist preacher's youngest and unmarried son, Wayne Huffman - a sergeant in the Army – tragically died in World War II Europe, in 1944. Older, married son William Guy Huffman, Jr. - a retail businessman - died in 1974. The pastor's widow, Floy - succumbing to her treated but spreading cancer - died in 1984. But back in April '41 she and her husband had another sudden death to deal with, one that hit Floy much harder than any alleged aliens. Her aging mother passed away (apparently in Lebanon, Missouri, across the state) on April 22nd. This would undoubtedly entail an emergency leave from the Red Star Tabernacle for the devoutly religious Huffmans and a long drive across the state to attend to the funeral and burial arrangements. So any follow-up visit by a nervous photographer to the Huffman home - as is alleged (see Chapter Two) - would had to have been on or before the 22nd, still somewhat fitting (barely) the "two weeks or so after" the UFO crash time-frame as recalled by granddaughter Charlette. Or, perhaps the shutterbug stopped by the house in time for the Huffman daughter-in-law's delivery of their first grandchild. Yes, the spring of '41 was the most eventful roller-coaster of both quiet William and Floy's entire life. A time of unforgettable, emotional, life-changing events. The boggling air crash episode happened "about three or four weeks before" the birth of Charlette's older sister on May 3rd.

What and *when* Minister Huffman's then-pregnant daughter-in-law was told by the family about the paranormal affair is unknown since she refuses to speak out publicly, other than briefly acknowledge that her younger sister Charlette had been telling the truth. Since William's daughter-in-law was just weeks away from giving birth at the time of the crash, Floy and her two fine sons may not have wished to disturb or upset her in any way with retelling the saga until after she had her child early the next month. She of course went on to produce a second healthy daughter (Charlette) later in the 1940s. As a sign that he successfully clammed up and kept up a normal, respectable routine, Reverend Huffman was hired as full-time minister by the Red Star Tabernacle on July 31st, 1941. As mentioned, William changed a bit, however. He used to be more "rigid," Floy told Charlette, but now was more open to the otherworldly and possibilities within God's mighty physical Creation.

Although the pastor and Floy pulled up stakes, sold their N. Main Street house - which they had bought from Red Star two years earlier - and left town in the fall of 1944. The loving middle-aged couple headed for a small Oklahoma village and church, initially. Their son Guy and his wife continued their young family life elsewhere in the Midwest, later settling in a small Kansas town. Their two girls were still not quite old enough and mature enough to handle all the bizarre details they may have overheard from Guy as the late '40s and early 1950s progressed. Still, the whispers and rumors little Charlette heard at dinner parties and family reunions over the years greatly intrigued her, quite naturally. But nothing did more than that aging copy of the unusual "3x4ish" b/w photograph taken at the scene, the little snapshot of the dead alien in "a rural field-like country setting," held up by the two plainclothes men (see Chapter Two). Young Charlette got a glimpse of this many times as the amazing picture was bandied about by older kin, until it was "lost" around 1953 or '54.

In 1984, Charlette interviewed ill Grandma Floy repeatedly, day after day, until the whole 1941 crash experience was relayed with as much detail as possible. Sadly, Floy finally passed away from her cancer later that year, and the family naturally mourned. Grieving Charlette Huffman Mann gathered up as many facts as she could, taking *years* to accumulate what she knew as she dealt with her own contemporary problems. Charlette finally typed up a letter summarizing the paranormal allegation in early 1991, then summoned up the courage to mail it - just weeks after the unobserved fiftieth anniversary of the event - to a respected UFO researcher named Ray Fowler. He felt it quite substantial and fascinating, and quickly did some investigating, but subsequently passed Charlette's story along to a more experienced UFO investigator/author named Leonard Stringfield. A respected former Air Force pilot and intelligence officer who had been digging into crashed alien technology for decades, Leonard visited and interviewed Charlette at length, repeatedly, and felt comfortable she was a woman of integrity and believability, telling the whole truth and nothing but the truth. Stringfield - who died in 1994 - was never able to catch Mann in any sort of lie or distortion, the family story remaining the same every time.

Mrs. Mann's case was further aided by her older sister, who while not wishing to get too involved also told Charlette what she had learned over the years, but was of course still in the womb at the time of the accident. While she preferred to remain anonymous and quiet, the sister did sign a notarized, sworn affidavit regarding the crash details, backing up Charlette's amazing tale. Researcher Stringfield was so impressed by the candid accounts he included the allegation in a 1991 UFO newsletter and within a book that year called *"UFO Crash/Retrievals, The Inner Sanctum, Status Report VI"* where it subsequently picked up at least a *little* more notoriety, some fifty years from the time of the original incident. The obscure but forthright book certainly raised some eyebrows within the hardcore UFO community. From there, even more regarded and serious researchers like Dr. Stanton Friedman and the father-son team of Robert and Ryan Wood reviewed the case very thoroughly, interviewed Charlette, and eventually reached Mr. Stringfield's conclusion: *the Huffman/Mann story is valid, unchanging, and reasonably strong, especially when backed up by some revealing government documents.* As researcher Ryan Wood summed up: "The first-ever-known UFO crash retrieval case occurred in 1941 in Cape Girardeau, Missouri," and this momentous event in turn apparently created the very first alien "reverse engineering work" by secretive United States military/governmental programs, making the story even more explosive. And *perhaps* also some startling "technological gains," as we shall see, quite *possibly* including the legendary, world-affecting "Manhattan Project." The Atom Bomb (see Chapters Two and Six), and possibly still other weaponry and technology, often guided by top scientist Dr. Vannevar Bush. The full impact of the claims is staggering, if true.

One thing is certain: the highest-ranking members of the U.S. military and the Roosevelt administration filed into the Oval Office for closed-door meetings with the president in the days after April 12th, including one on the 16th that was logged in the record book as "*off*." This meant the subject matter was so explosive no notes were allowed to be taken; it was considered very classified and "*off the record*." Harry Hopkins and George C. Marshall attended at least two of these White House meetings with FDR, along with Secretary of State Cordell Hull. Another name that popped up in the Oval Office logs in this period was the mentioned man of science and technology, trusty Vannevar Bush, a very significant sign for this saga as we shall see later.

"MO41" now seems to be a handy term for not just the 1941 Missouri crash and its Huffman-involved response, but also the secretive Army recovery and government research of the alien materials involved. It is slang, perhaps, but encompassing "MO41" should rank someday right up there with "Roswell" in mainstream terminology for real-life "UFO crash" information. The sensational MO41 nonfiction saga is nowadays even more open for review than ever, having been included within the chapters of several books, a magazine article, and a 2003 SyFy Channel television documentary on alien crashes. Later it was touched upon in a 2014 History Channel TV program called "*Hangar 1: Alien Technology*." And it has been explored on various online web sites, within some newspaper reports, on national radio programs and podcasts, and at serious UFO conference presentations. These sources, along with several others, all believe the remarkable 1941 southeastern Missouri alien crash allegation is a truthful and candid claim, thanks partly, or mostly, to Charlette's retelling of what the preacher saw.

What the youngest Huffman grandchild witnessed on her own is almost as important to more fully understanding MO41, the astounding bombshell well before Roswell.

CHAPTER TWO

What the Granddaughter Saw

"You couldn't see those eyes and not be affected."

In mid-1991, Tyler, Texas-based Charlette Mann broke the
Huffman family's general silence on their alien crash
shocker, respectfully waiting several years after
grandmother Floy Huffman's death from cancer. {And
Charlette's still there, to this day.} In that interim, Mrs.
Mann spoke to other relatives, gathered more information,
got the facts straight, and set "the family secret" down in a
coherent letter, typed up to send to the right UFO
researcher. She eventually selected an investigator named
Raymond Fowler, and he soon began digging into the tale,
mostly at first with his friend, researcher Leonard
Stringfield.

Much of the Huffman saga we've now learned, but a
remaining large part of the overall original story is what
was hand-delivered to the caring Cape Girardeau
clergyman an estimated two weeks or so after the crash
incident, perhaps just prior to April 22nd, 1941 (when Floy
Huffman's mother died). It was startling proof - *physical
hard proof* - and perhaps the one last scrap of evidence
from that night that the American military did *not* manage
to get their mitts on - at least not initially. It became
something the Huffman brood - Charlette most memorably
- set their eyes on. And then their memory banks.

According to Charlette Huffman Mann, one of at least two people "that night" with cameras blazing at the MO41 crash scene caught up one day with Reverend Huffman at his Cape Girardeau house on North Main Street, just up the boulevard from the big wooden Red Star house of worship. The unnamed visitor first peered around the street and sidewalks with great caution. "He looked very, very frightened," was how Floy recalled the visitor's distinct expression and behavior as he arrived on the cool, concrete front porch (according to Charlette's retelling). That's how effective the United States Army was at browbeating and warning those on the scene to keep quiet, perhaps reinforcing this in the days afterwards as well. Possibly the local FBI and/or Cape police joined in the admonishing too. The cautious, protective military - and likely the federal government - wanted this wild case kept as silent as possible.

By late April 1941, the shutterbug visitor to the Huffman's rented house was naturally worried about being seen or overheard with someone who was "there" that night, but he had a gift, of sorts, to impart. It was a hot potato. A surviving smoking gun photograph of that unforgettable, traumatic night.

In looking back, some MO41 researchers feel it is quite possible that the unnamed male visitor to the nearly-riverside Huffman home was the tall, crusty, cigar-chomping reporter/photographer for the local newspaper, *The Southeast Missourian*, Garland D. Fronabarger. As mentioned, "Frony" was his nickname, and his birthday, interestingly enough, fell on a day close to the timing of the alien crash and its aftermath, April 24th (1904). It should be noted that clean-cut but blunt Fronabarger shared the same religious beliefs and interest in beautiful roses as Huffman, thus the two men *may* have been friends *before* MO41 went down. Both men were former school teachers in small Midwestern towns before they found their true calling in life. While a fellow Baptist, G. D. Fronabarger was *not* a member of the Red Star congregation, however, but he at times frequented Cape Girardeau's Capaha Park, where a special rose garden was situated and can still be frequented today. It is where *both* men may well have visited that very springtime, since both loved roses. Cape was known, after all, as *"The City of Roses."* They made good photographic subjects, these glorious, colorful blooms. {Ironically, the Rose Garden at the White House was also of interest to America's First Couple, since "Roos" in the Dutch language within "Roosevelt" meant "rose," their favorite flower.}

G.D. Fronabarger became a cub reporter for *The Southeast Missourian* in 1927, took up photography in 1929, and was hired full time in 1931. A pair of newspaper employees who knew him stated he was seldom without a smoke, a notepad and pen, a big camera, and his familiar brimmed hat. The same 2014 e-mail source that described nurses "white with fright" as they stared at the dead alien forms that spooky night also assured that it was none other than Garland D. Fronabarger at the scene, plying his trade. At least that is the way his sister told the story, as heard from a friend in the know. With his trademark attire and gear, tall Frony cut the most imposing and memorable image at the scene and in later recounts of it by the sibling, who told the tale right from the Spring of '41, for decades forward. This account was backed up by a second relative of this Cape Girardeau woman; that Fronabarger was in fact the only person identified by name, since he memorably interviewed people at the site and likely introduced himself as a reporter, in order to get the facts in this exciting, seemingly historic story.

In addition to news writing and reporting, Frony took up photography in the Great Depression, somewhat for fun but mostly to add to his professional value and income. In the early 1930s he purchased and used almost daily for many decades a large Graflex "Speed Graphic" camera, a rather heavy beauty with its glass lens extended out at the end of flexible bellows, often poised atop a metal tripod. He also acquired a "4-A Special Kodak" from his local photography mentor, Chester Kassel. Fronabarger often developed rolls of film or copy photographic exposures (from slide film) when working "down in the furnace room" at his newspaper's offices (later moved upstairs), or in a darkroom at home, on South Park Street in Cape. He grew quite capable in all aspects of his pastime. By all accounts, Garland was a very respected and popular man in his community, reporting on stories for the paper for over six decades, often without credit or name in any bylines. He was a dedicated husband and a father to a teenaged son in 1941. Frony's picture-taking skills became so advanced he utilized them outside the confines of professional journalism and at times sold private copies of his fine work. All the while, the rather irascible news hawk always enjoyed smoking big cigars, even when developing his photos in a dark room. Besides horticulture - at least rose-growing - Garland was also fond of area hunting and fishing in his spare time.

Ubiquitous G. D. Fronabarger made friends with many in southeastern Missouri, including a visiting judge in the early 1930s, a Kansas City-area man named Harry S Truman (long before he gained great political power). But after he retired from the local newspaper in 1986 Garland's health began to fail. Aging Frony - greatly prompted by his second wife - held an auction of his possessions, including many of his photographs. Some were sold, some were tossed out, and others were left for a private collection currently held by his old employers at the *Missourian*. Some were also saved by his son, John, who revealed in 2013 he subsequently lost many of them. While it is true that not a single one of the remaining pictures or negatives today show a crash scene, spaceship, alien bodies, or debris, these are only a small portion of the entire once-great aggregation. Asked whether his father could have kept the life-changing story of an alien crash to himself, John replied that his dad "was talkative about some things, and secretive about other things." So that yes, it was at least *possible*.

One great overlooked factor in the photo mystery is that the image was a very awkward posing for the carefully-arranged snapshot. That is, the unnamed photographer did not walk up to an alien creature laying on the ground that night and impulsively snap its picture, perhaps with someone caught in the act of kneeling over it, like William Huffman himself. No, the old b/w photo Charlette and her sister saw was an evocatively *staged* photographic effort - Frony's trademark. Posing was a professional trait that earned G.D. Fronabarger some small amount of ribbing; he didn't take too many "action photos" of people. Instead, Garland had his human subjects stop and hold a rather stiff pose, sometimes rather awkwardly, urging them not to move or blink as he readied his camera and snapped the shutter. The two unknown men shown in the Huffman family snapshot were posing just like that in a farmyard, holding the gray extraterrestrial up between them, as witnessed by Reverend Huffman at the scene. It was not a "candid camera shot," but quite obviously arranged. This would have been a classic Frony approach.

However, there is one possible problem with busy newsman G. D. Fronabarger being the anxious source who handed over the special picture at the Huffman home, in the accident's wake. As mentioned earlier, Garland usually developed his own photos, mostly at work or at home, and *his snapshots were always trimmed with a brief, straight, white border.* The unique photograph that the visitor handed over to the reverend that late April 1941 featured *scalloped* white edges. That kind of dissimilar edging was applied from somewhat special developing paper, often associated with public snapshot developers, such as at a drug store or camera shop or commercial photo lab. Frony's developed professional Graflex prints - still available for review to this day - were all in the 4x5 or 5x7 inch size. Charlette Mann memories of the posed MO41 photo reveal it was about "3x4-ish" in size, give or take a little.

{A 1999 interview with Charlette was published online stated that Mrs. Mann recalled the eerie ET picture as "approximately nine inches by nine inches." This does not appear to be an accurate printing of her statement, an apparent typo.}

Still, we have a possible modus operandi for Garland to be the photographer at the scene and yet have his own snapshot come out two weeks later, developed for him by an outsider in an unusual (for him) size and border edging. He might have sent the small camera's roll to St. Louis, to be developed. Where exactly could this outside process have taken place? A commercial photo lab in St. Louis called "Fojo" commonly developed camera film rolls for Cape Girardeau citizens who mailed them in, back in the 1930s and '40s (and beyond). *Mailing* a roll of film to a St. Louis firm for professional but different development styles would take about a week to ten days to get it returned, fitting our post-crash time frame well, as we shall see. But why send anything off "to be made up" when there was competent Fronabarger working down in his basement photo labs (at work and at home); his pals Kassel and Haman working a few blocks away from Frony's office; and still other photographic development sources in town to be tapped? In the MO41 case, an outside development firm might have been *purposely* used "while the heat died down" in Cape Girardeau. To deliberately get the evidence out of town for a while, so to speak. Especially if a newsman's office or darkroom or even *home* was being searched by authorities. However, this is simply a theory.

According to Mrs. Mann, the unexpected visitor to William and Floy's 1941 Cape Girardeau home commented tellingly: "I have a copy, and I want you to have a copy, for safekeeping." So apparently multiple copies were carefully made up with great care, perhaps by a real "camera bug," so to speak, "about two weeks after" the crash. Therefore, the photographer who delivered the snapshot to William may well have had to wait like everyone else while his roll of film was developed and finally mailed back to him. Or... the cameraman had copies made up by Chester Kassel, who operated his photography business on Main Street, not far from the Huffman dwelling.

If it was truly Fronabarger on the Huffman front stoop, he was acting a little disobedient and rebellious by giving the friendly pastor a copy of his small personal camera's production. Thus G.D. could somewhat avenge the Army's rude nighttime takeaway around two weeks earlier; it was carefully entrusted from one Cape Girardeau Baptist to another.

{Another mark in favor of Frony being the unnamed visitor on the Huffman front steps was that Fronabarger was a newsman who was familiar with methods to efficiently track down local citizens at home or work, in getting quotes and facts for newspaper stories, and/or for snapping posed photos. As someone fairly new to Cape, the William Huffman name in a church-rental home likely did not show up in a 1941 city directory, but Garland knew the tricks of the trade in sniffing out sources, possibly stopping by Red Star Tabernacle first to ask about the preacher's home address. It is also plausible that perhaps across-the-street neighbor Milton Cobb was also given a photo copy, and that this was another reason why the unnamed photo buff made the trip; he knew it was a virtual "two-for-one special" in that both men lived so close by on North Main Street.}

If the person at the crash scene had taken *more* than one rather illicit photo of the alien tragedy with the small pocket camera, then where were all *those* resulting poses and prints? In Reverend Huffman's late night recollection to his family, the pastor watched the two local men prop up the strange, limp creature that April night for the photographer's careful snapshot, and seemingly it wasn't too long after that when the army arrived. Thus it seems very possible that the posed ET image was the *only* print on the small hobby camera's roll of film that resulted from the incredible accident. And what was Garland Fronabarger's nickname? "One Shot Frony." *His reputation was such that he often just took one snap when people posed and then walked away.* Could that have been the case here with the small pocket camera and its posed ET picture? With his larger, work-related "main camera" - the rather klunky "Speed Graphic" - Frony likely took *many* crash images, mostly posed carefully one would imagine, but perhaps some in full glorious color, since the staggering farm scene was truly historic and the newspaper was paying for his film and flashbulbs during tough economic times. But with his *own* hobby camera and some cheaper, black-and-white film. well. G. D. Fronabarger was known by friends to be a bit of a cheapskate, and thus we can see why he would have shown up with just the *one* souvenir when standing on Huffman's front porch one fine spring day. {Plus, the small pocketed camera might simply have been low on film or flash bulbs at the time, too.}

Whomever it was visiting the Huffman home on North Main Street said as mentioned that he had kept at least one duplicate (maybe more?) of the dazzling little photograph for safekeeping, this posed snapshot with the scalloped sides. "I think you need to have this," the visitor had told the pastor. Lost to the world to this day, the startling photographic evidence is the "Holy Grail" from the amazing saga, searched for by intrepid followers of the haunting case, always tantalizingly out of reach. The picture's value in today's world would of course be priceless, especially if the negative was with it, to back it up authenticity.

As the visitor observed, startled William Huffman stared down at the unusual black-and-white image snapped of the two local gentlemen posing at the local farm's MO41 site, holding up a dead gray alien, its long, floppy arms outstretched against the warm night air. A flood of memories likely came roaring through the pastor's mind. He must have wondered to himself what to do. Should he give it back? Should he destroy it? This was a fascinating relic, a smoking gun, and it excited him - and yet likely terrified him at the same time. It was pure gold, yet also a dangerous, ticking time bomb, since the surveillance-happy FBI was in town, several men of the community were in the Army reserve, and many citizens were capable of spilling secrets. No one wanted to be left "holding the bag," and that was probably one reason why the cameraman wanted to get rid of it in the first place, yet he could not bring himself to burn it or toss it out in the trash.

William probably shuddered and tried to block the darker memories from that night – the burning wreck site, the weird rubbery dead bodies, the army, the threats - from his mind. Yet he didn't say "no thanks" and refuse to accept what was offered to him.

The photographer actually *asked* if Reverend Huffman would take it off his hands, but his demeanor and approach were such as to be obvious in his need of psychological relief. The visitor required the answer to be yes, very badly. "You're the only one I trust with it," the cameraman supposedly claimed, possibly indicating that he knew the evangelist pretty well and - if this was an honest statement - that the visitor was *not* going to go handing out such souvenirs to others from the crash scene, nor anyone else in town.

William grudgingly accepted this controversial "gift," likely with a sigh. The striking, memorable picture was now Reverend Huffman's possession and problem. The visitor likely made some small talk as William dealt with the bizarre reverie, then thanked his evangelical friend, bid him adieu, and took off down the sidewalk. Quite possibly they'd see each other socially around town now and then.

At one point, Pastor Huffman was so nervous about the startling ET snapshot he soon gave it to his oldest son Guy, for safekeeping. In this way, the anxious preacher/fund-raiser would have his own plausible deniability if the military or any U.S. government agent ever came snooping to discover if he possessed any remaining evidence. William knew that Guy knew what it was all about. The picture would remain in the Huffman's cozy house for a while, just not in William's possession. Maybe William knew that Guy would be moving away soon. Guy was more worried about another prize in the Huffman home, ready to change his life: his unborn daughter, now just days away from birth in a Cape hospital. Perhaps Guy now and then quietly showed the cadaver snapshot to a few relatives or close, trusted friends, but mostly just tucked it away somewhere secure and got on with life.

This would have been the end of this story had the exposure not become a private curiosity, carefully passed around later at Guy's home on special family occasions. {Which means that Guy and his wife and perhaps others occasionally discussed MO41, but not his parents.} Charlette says she saw it "about twenty times." Mostly furtive glimpses, sometimes more at length. It spooked her, yet intrigued her, every time. Chills likely went down her spine and the hair may well have raised on her neck. Photos from magazines or newspapers, or movie stills, didn't show this kind of "sci-fi" subject in those days, real or faked. The memories haunted her for the rest of her life, since young Charlette's father generally kept it out of her reach, making it even *more* sensational, enticing, and mysterious to her. Whenever Kansas furniture store salesman Guy Huffman showed it to a select number of trusted adults in the late 1940s and early 1950s, little Charlette would move in closer to sneak peeks and listen to small snippets of her daddy's tantalizing tales about child-sized "space creatures" found by grandpa at the scene of a terrible accident on a farm near some distant place called "Cape Girardeau, Missouri." Generally, Guy kept the picture hidden away so it wouldn't get lost, damaged, or stolen, a policy he learned to his great regret he should have stuck with as the next few years rolled by and MO41 became a distant, strange memory.

Decades later, Charlette Huffman Mann described to contemporary investigators the shocking ET photograph in every detail possible. It did not show color but appeared to be aged by a few whitish lines in the print, as if accidently folded or dog-eared at times. It was surrounded as mentioned on all four sides by the white scalloped or rough edging, which helped set it apart from most b/w exposures of the day. But it's what was *in* the picture, the subject matter, that was the most eye-catching feature of all.

At the bottom of the unique, treasured photo, in the foreground, were visible wisps of grass or weeds, just a few inches high in most places. This likely indicates it was taken in someone's *yard or nearby unplowed field*, not way out in a sewn pasture. "Some trees" were visible in the background, one of them fairly large, William's granddaughter added. Overall, the spine-tingling snapshot had a "rural field-like country setting" to it, Charlette described decades later to a researcher.

Two unidentified bland Caucasian males held up the dead extraterrestrial for a proud pose in the picture's center, perhaps needing to appear along with the creature partly "for scale" and partly "for grins." Charlette never learned who the two humans were, or why they wanted to proudly pose with the odd dead "critter" in such a way. The creepy cadaver was propped up by the average-sized men gripping it in the underarm area, and then holding their human arms out to support and extend the alien's unusually long, slender arms, which outdistanced their own. The unidentified man on the right side of the snapshot was wearing dark pants, a white collared shirt with the sleeves rolled up (once again indicating that it was a fairly warm - "but not hot" - evening), and a wide-brimmed hat. The unknown male on the picture's left side had on slacks, a white collared shirt, and "a short waistcoat jacket." There were no other people (or aliens) in the foreground or background of the creepy shot.

In a rather crude, copyrighted, stick-figure drawing she made for a UFO researcher in 1999, Charlette recreated on paper two human figures - *both* wearing brimmed hats - holding up the dead alien with two hands apiece, each gripping the lifeless ET's wrist and underarm areas. The grown men tower over the creature, which displayed the same features the Huffman granddaughter described years earlier in interviews, with no changes, indicating again her consistency in retelling the original story.

The 1941 photo's extraterrestrial was thin and soft-looking, almost as if "he had no bone structure." It was sporting either a very rumpled but tight-fitting flight suit or his own unusual skin. Either way it was very eye-catching, appearing to Charlette like "crinkled, soft aluminum foil." It was difficult to tell, *suit or skin?* Could this outer body visual appearance in the black-and-white photograph have been a protective thin layer of metallic fabric in an artificial body covering for space travel? Or just the appearance of the nonhuman creature's delicate skin after a traumatic, sudden physical death - perhaps a painful one - when propped up by the two men in the picture? Or some sort of strange combination of *both*? "It looked like it did not have clothes on," was about all Mrs. Mann could sum up at one point. No zippers or patterns, seams or buttons were visible, but the outer layer of the creature was definitely "shiny" in appearance.

It is of course very possible that the ET in question was never prepared or *supposed* to be exposed to this planet's alien atmosphere, and when "he" (for lack of a better term) did, his skin wrinkled badly, either before or after his sudden, tragic demise. If his "heart" had stopped beating, and thus blood circulating, the alien's epidermis could conceivably have been gravely affected - if it even had such organs. Before death set in his epidermal layer was perhaps also altered by exposure to earth's air pressure, temperature, and moisture level, and after death the same, and perhaps still other changes *within* its body, and without (outside the comforts of its climate controlled ship). Whatever the reason, this "crinkly aluminum foil effect" was a noticeable condition that was not just visible on the body, but atop the entity's large, perhaps otherwise bald head. William Huffman really didn't go into this area or level of detail to his family in '41, evidently.

Another factor to consider of the "wrinkled foil" description is that simply the light from the photographer's flashbulb and possible mild movement of the subject by the two local men in posing him may have caused the dead alien's image to blur slightly, fooling little Charlette's memory a bit. The wrinkly look also might have even been an emotional or psychological reaction by the terrified creature before death. William Huffman was busy kneeling over a limp ET that was barely moving his chest, unable to breathe well. It could well be, literally, *the air didn't agree with him.* It might have wrinkled or even shrunk a bit his otherworldly body, and its content might have destroyed his respiratory system, and thus his entire life. The same goes for all three identical-looking beings.

But *who* exactly was posing in the strange picture? Is it possible that at least one of the two men clutching the ET was either Cape Girardeau County Coroner, Narval B. Short, or a man working for him? Norval was in charge of handling dead bodies as a daily part of his job and thus was the most likely man at the site to have been allowed by the police and fire departments to touch and examine - and perhaps prop up - the deceased alien creatures. It is entirely plausible that another county's coroner was called in and was in the snapshot as well. Or perhaps men who worked for them. Or a mortician, funeral home director, or undertaker, all men accustomed to handling badly affected corpses. Cape County's official Medical Examiner, Dr. J. H. Cochran, might also have been called in that night, along with the previously mentioned Justice of the Peace Milton Cobb. Yet for Cobb, Short, and Cochran, or any other professional man, to have yanked up and toyed with a dead body - even a very foreign corpse like this one - was rather undignified behavior and potentially bad for their practiced reputations in town. Thus their participation would have been fairly unlikely. The same could be stated for any police officer, fireman, federal agent, or military personnel in civvies at the scene that night.

Any male in the area that day might have been wearing a hat as described, as that was the style at the time, but at night? He might also have had his sleeves rolled up as it was so darned unseasonably *warm* by that late afternoon, but once again, at night? It had cooled a bit. Whatever the case is, the two men posing were not in uniforms, so they were unlikely to be cops or firemen or soldiers, unless they were off–duty and perhaps in a silly (or inebriated) mood.

It would also seem logical also to conclude that whomever it was that posed with the wilted, unconventional creature would have first asked the photographer - newsman Fronabarger? - to have taken their picture special. Or was it vice-versa? If the photographer whipped out his camera that night, supposedly from a shirt pocket, then wouldn't *he* have asked men to pose with the shocking, dead visitor? It should be noted that the Huffman granddaughter did not recall seeing the two photo men grinning or clowning in any way with the creepy body, just holding it up and out. And holding still. But *someone* wanted the picture snapped on purpose. Precisely whom it was that posed, that may always remain a mystery.

If *any* local news reporter was indeed the photographer of the special ET snapshot, his thinking at the posed crash scene might have been: "*I've got to record this scene for posterity, for the newspaper, and for my personal collection. It might make me a lot money, or even famous someday.*" But Charlette said that there were. *two* photographers present - the other completely unknown; was it area cameramen Chester Kassel or Jim Haman? - took so many snaps of the scene, perhaps they ran out of rolls of film and/or flashbulbs in their larger cameras. It happens. Neither man was expecting to be called into the most sensational story in human history, not on a slow Saturday night at the end of a long week. No, instead one fellow whipped out his backup, his little personal pocket camera, kept with him for just such impromptu moments or emergencies.

Initially, in any news hound's mind the amazing remaining smaller snapshot might soon be splashed along with other Speed Graphic photos across the pages of the next *Southeast Missourian* as a blockbuster scoop, the envy of all journalists everywhere. At the least, this crash oddity would run as a novelty story, perhaps picked up by the wire news services, and go national. He'd be world famous soon!

In the Charlette Mann-described crash scene exposure, the notable man in the wide-brimmed, outdoorsy hat, with his sleeves rolled up, that clue could well indicate it was the crash farm's *owner*, posturing with a willing pal or his kin. A classic *"Looky-what-I-found"* picture. True, a country farmer usually wore overalls, not "white collared shirts," but this may well have been a Saturday night, when so many country folks get cleaned up, don something more respectable, and get ready to go out on the town.

The two photo subjects were needed by any experienced newsman to carefully pose with their new and unique prize in the snapshot, for scale and background, to best fully tell the story in one easily-grasped setting, all of which a newsperson likely expected to have a big impact come Monday afternoon's newspaper deliveries. {Saturday evening's paper was already out and there were no *Missourian* Sunday editions printed in 1941; Monday's edition would feature mostly Easter Sunday photos of churchgoers.} "Prop that scrawny thing up and hold still," one can imagine crusty "One Shot Frony" barking as he aimed his little backup camera, stepping forward in the soft young grass. Yes, Garland pressing the shutter on his smaller camera for his two human and one nonhuman subjects that night would essentially be the same situation as his going to a 4-H Club or local district fair and having an area farmer pose with his prize pig.

The early 1950s' little girl named Charlette who peered in at the stunning picture over a decade after its creation and duplication was quite mesmerized by its unique subject matter, mostly its two very large, dark black, oval-shaped eyes. The silver-grayish "child-sized" person with floppy, skinny arms and legs, and slim torso, featured a very large, rounded, hairless head. This balloonish, cartoonish cranium seemed too big for its petite, crinkly body. That was eerie even for the transfixed Huffman adults who stared at the photographic image, but for a young lady it was perhaps the stuff of nightmares. Yet. she couldn't get enough. Charlette longed to see it at length and ask questions, likely her older sister too. Certainly every time their dad got the picture out to show off, and they caught glimpses, the image of the weird creature shook them. But the explosive subject was not up for discussion, not for a little girl in such a conservative era.

Charlette Huffman Mann remembered in the 1990s that the wrinkly ET cadaver's arms were "very much longer than our arms." Charlette was also haunted by its creepy lack of a nose or human-sized nostrils, "just two little dots" below the eyes, and above its thin, small "mouth." Charlette clearly recalled the creature's lack of a visible human-like oral cavity, or even lips. Just "a slit across" the lower part of its face, it seemed. And ears? It seemingly had none, nor lobes or even openings on the sides of its head, at least not that she could make out, or recall years later.

The photo ET seemed expressionless, evidently lifeless, and flaccid, yet appeared to be a very real entity, not an elaborate doll, not a silly hoax. It clearly wasn't of this earth. This fits with the low-tones rumors and references to the sudden otherworldly wreck that Charlette overheard her shopkeeper father mention briefly during some family get-togethers in the 1950s, and it also fit with what dying Floy Huffman revealed in the 1980s of William's account from

April of '41. This picture was fantastic *history*, but *hidden* history.

The late extraterrestrial astronaut's hands in the snapshot were also of special interest, from Charlette's point of view, both as a child and as an adult. The oddly long arms ended in strikingly long fingers, and apparently featured only three per hand, along with a long thumb. There were no visible fingernails or thumbnails. No jewelry or tattoos or markings of any kind. The unforgettable critter was strangely undamaged - outwardly - after the horrific crash. That is, it had no *visible* injuries, such as a bleeding wound, protruding bones, or dark bruises. Or even torn clothing or burn marks or holes, confirming what her grandfather stated. Other than two long, skinny legs winding down into the grass, Charlette could not recall if the photo creature had any feet, if even visible slightly in the photo. They were possibly just out of shot. But the posed extraterrestrial sure had some spooky appendages, overall. Just not quite human, but fairly close. Certainly it was also not reported that the aliens once had any helmets or headgear, multi-faceted or pocketed spacesuits, and special breathing apparatus to handle our atmosphere, strongly indicating that they were not intending to be outside of their ship, on terra firma. They had no buttons or bows, jewelry or gloves. The more one ponders their attire, the stranger it seems.

In the 1990's, Charlette told researchers she specifically remembered that her Gramma Floy repeated that Grampa William stated he was still kneeling and praying over the last living humanoid - who was taking short breaths - *as* the men were posing for the photographer with one of the two other dead "creatures." Apparently William was just outside the shot, then, making sure *he* did not to touch the unusual bodies. Evidently there was no sign of shock, partial consciousness, or life itself in the alien in the old b/w picture that the little Huffman granddaughter stared at a decade later. "It" or "he" just seemed like her grandpa said: about four feet tall at most, soft and thin, gray and lifeless. But was it a "he" or a "she," or just an artificial clone? Whether any of these three not-so-human "astronauts" exhibited any sign of gender or sexuality - male, female, or whatever - it was not stated, but was likely not visible at all. Were these aliens actually lifelike robots or computer remote-controlled androids? With no special markings, growths, deformities, or piercings, or really *anything* that made them stand apart from each other, these particular voiceless extraterrestrials were strangely identical and bland to the point of soullessness, as if they came out of a cookie cutter.

What of the ET auditory systems, if any? Could they hear? Could the still-living creature understand the English language and what was being said by the crowd near him that night? Reverend Huffman was somewhat of a latecomer to the scene that night, leading one to wonder if any of the three creatures spoke or communicated in any way prior to his arrival. They may have been on the ground for up to an hour, maybe two, prior to the cop and the fund-raising pastor's presence at the crash scene. It is possible open contact was made but simply never mentioned to the evangelist. He evidently never heard a word from the beings, or even so much as a grunt, nor did he report anyone telling him so in the aftermath of their deaths.

Other questions come to mind. Were the ETs able to fully grasp the gravity - so to speak - of their situation? Were they in an altered state, perhaps drugged or drunk, or dazed and confused during and after the dire, traumatic accident? Or were they able to understand everything, and even human thought telepathically? When alive and well or when dying, could they somehow communicate their own thoughts without the need for vocal cords? We have no evidence or allegation for this, but other claimed alien encounters over the past few decades would state this. Did the crash kill the two aliens and result in the third one's death, or were they dead in the air, well before they hit the ground? Obviously, the unique "visitors'" time on earth was simply too tantalizingly brief to draw any sort of firm conclusions.

When she thought back to the limp MO41 alien's "slightly-slanted" eyes in the posed picture, Charlette Mann remembered they took up much of the alien's face, such as it was. But were there any eyelids? Eyelashes? Pupils? Iris? Did the aliens blink? Charlette could not tell from the picture, and the family patriarch William never described such detailed things. But those large, dark, somewhat almond-shaped eyes! They were unforgettably impassive and cold, inky black and unexpressive. To perhaps complicate matters, snapping the ET's picture required flash bulbs that would pop loudly and glare when pictures were recorded in those days. Usually the flash of photos of human beings can reflect light in the eyeballs of the photo subjects, and make us blink, but the creature? Apparently his eyes were simply blank and black, with no shine or sign of life as we know it. Ironically, the big glass flashbulbs from cameras of the 1940s were often hot and cumbersome to handle when taking photographs anywhere; now ironically the resulting picture would be just the same way for the Baptist family. But when little Charlette saw the stupefying photograph, she admitted, she was captivated most by the black emptiness of those huge, creepy alien eyes. That sight haunted her for the rest of her life, quite naturally. "You couldn't see those eyes and not be affected," Mrs. Mann aptly recalled in an on-camera interview. They dominated the creature's skull and thoroughly spooked the photograph's observer, for good.

One clue as to the validity of what Charlette Huffman Mann claims on the alien she saw comes in the form of a leaked 1954 U.S. Army manual. This once-classified *"Special Operations Manual"* for alien artifact recovery by an elite American military unit covers vivid descriptions of "Extraterrestrial Biological Entities" hauled in previously and studied. One of two types is explained as "humanoid," but with "a head much larger than humans." The eyes of this species of creature were described as "very large, slanted, and nearly wrapped around the side of the skull. They are black with no whites showing." This particular alien race's nose "consists of two small slits which sit high above a slit-like mouth," with no external ears. There's no hair on their bluish-gray skin, with two remarkably long, slim arms and legs, the '54 document claims. Both hands feature "three long tapering fingers and a thumb." Without ever seeing this stunningly match-filled description, unwitting Mrs. Mann's recall of the photographed alien cadaver becomes even more impressive. Point by point, the United States Army Intelligence leadership secretly printed up a document that privately told the same exact tale a half century earlier than the Mann public revelations in the decade of the 1990s (*before* the S.O.M. was revealed to the public) and re-interviewed at length just after the turn of the new century. The manual appears to be authentic and very detailed, available for viewing at MajesticDocuments.com.

Although not wishing to get involved publicly in the MO41 story, Charlette's sister - growing nearly nine months large in her mother's womb during the incident - agreed with Mrs. Mann on the truth in the Huffman family saga. The older sister quietly attested to her sibling's statements in a private interview with a researcher. She supposedly confirmed everything her younger sister claimed about the photo and the 1941 explanation. She even signed a sworn affidavit on Charlette's Floy Huffman 1984 recollections and on the visual components of the resulting "family photo" she at times witnessed as a young child too. Sadly, Guy Huffman and his wife, and his brother Wayne, along with William and Floy, are all long since deceased, so they can't help us, evidently leaving no written testimonials behind. Even the Number One Suspect for the photography involved, Garland D. Fronabarger, has long since passed away. But there was one man known to the Huffman family who could have been of assistance, to the Huffmans themselves and to all historians and researchers, cosmologists and seekers of the truth. He was interviewed many years later on the unusual topic and appeared to be at least somewhat knowledgeable but very evasive. And this man's name was Walter Wayne Fisk.

Mr. Fisk was truly a man of mystery, someone who supposedly got his hands on the historic photograph, when Charlette was about eleven years old (in 1953), and frustratingly, the proof was never seen again. Charlette asserted Walter outright took it, and *he* claimed he gave it back. Who was telling the truth. and just who was this guy, Fisk? Why did he never return the Holy Grail of ET photographic evidence?

Revealing facts and mysterious claims about the key MO41 mystery man only came to light decades later, casting a bit of helpful light on a soul who evidently lived perhaps not quite in the shadows, but shielded his true identity as he worked in the light of day alongside the average Joe in a small Midwestern town, and later amongst the most powerful people on earth in our nation's capitol. He was apparently from the world of military "spooks," missions and misdirections.

"He was a friend of the family," that was how Charlette Mann recalled hearing from her parents about Mr. Walter Wayne Fisk, but would a true "friend" really take off with the small MO41 extraterrestrial photographic proof and never return it? According to Mrs. Mann, Fisk told the Huffmans way back in 1953 (or '54) that he dabbled in photography, and when he heard about the special snapshot, he quickly asked to see it for a close scrutiny. Highly intrigued, Walter then requested a favor: to take the black-and-white snapshot with him, supposedly to "authenticate it" and allegedly to show it off to his parents. Then he'd bring it right back, he assured Guy.

Unhappily for our story, Charlette's trusting father allowed his small town pal to walk away with the picture of all pictures. None of the Huffmans ever saw it again. Proof vanished overnight. After this unfortunate handoff, either the Huffmans "moved away quickly" or Fisk did, Charlette cannot quite recall now from her youth now. If it was Fisk who moved, he sure pulled a fast one in his getaway, managing to escape the Huffman brood's grasp and wrath for the rest of his life, or at least until near the very end.

To start out and sort out this sticky situation, we must first understand that Guy Huffman and Walter Fisk in the early 1950s were friends in a humble Kansas town, where the twosome often chatted on Main Street and at the local Chamber of Commerce, according to Charlette. At the time Walter was about thirty-one years old and Guy was around thirty-seven. According to a difficult interview Linda Wallace once tried to conduct with an elderly, curmudgeonly Walter W. Fisk, Guy worked in a furniture store on one side of Main Street, while Walter was situated in the insurance office on the other side. Since it was a small town in Kansas - probably Pratt, Fisk's hometown - everyone seemed likely to know everyone else's business. Sleepy and slow-paced were likely most apt descriptions of daily existence, hence folks struck up conversations and visits to each other's homes and businesses, partly just to kill boredom. At least, that is what Walter and Charlette Mann's father did.

The two men were enjoying a Huffman family dinner party one evening - Walter brought pheasant, Charlette vividly remembered - when the discussion moved to the possibility of intelligent life on other planets. Since 1947 "UFOs" and "flying saucer" reports were quite the topic of news stories, some far-out Hollywood movies, and casual conversation, now and then. During the course of this party chat, Guy of course remembered the posed 1941 ET photograph from Cape Girardeau he had quietly held onto, and his father's one-time wild alien crash scene saga, still strikingly imprinted in his mind. Guy made the decision to inform the insurance agent - considered "a close friend" - only of the spooky snapshot, which he then brought out to display to Walter. The two anonymous men holding up a lifeless gray alien on a farm, a simple three inch by four inch (or so) picture with white scalloped edges. But W. Guy Huffman II supposedly did *not* tell Walter W. Fisk about the MO41 crash back-story behind the alien picture, now a bit dog-eared after a dozen years. Guy kept his promise to his father William *not* to discuss that explosive Cape Girardeau angle, and only informed Walter that the propped-up photo image was of "a creature from another world." Fisk claimed to know plenty about photography and how to develop images, and was very interested in the Cape Girardeau snapshot. He said he wanted a closer look at it in his photo lab at home, and to show the exposure off to his parents. And with that, the mysterious Mr. Fisk nearly waltzed off into history. Nearly a half century passed in silence, as Charlette's maturity and yet also indignation grew.

Flash forward to the early 1990s, when noted UFO author and dogged investigator Dr. Stanton Friedman tracked down an elderly Walter Fisk, finding him in an attractive middle-class home in Albuquerque, New Mexico. Stanton called Walter and they chatted, but the researcher got little from his aging source. Fisk was cagey, even in his old age.

He seemed adept at dodging straight answers to direct questions. He claimed not to know anything about the valuable Huffman photo, and that he did not possess it. He did make some pretty bold claims otherwise, including that he had been an advisor to past U.S. presidents, and that he held a degree in psychology, once working as a psychologist. Stanton left, not sure what to think.

Not too long after this frustrating call by Friedman, investigative author Ryan Wood made a personal visit to the Fisk home on Mountain Road in New Mexico's largest city, very close to the local airport, Kirtland Air Force Base, Sandia National Laboratories, and the Manzano Storage Area, a place rumored to have at one time held closely-guarded nuclear weapons. and still other classified, secret, materials rumored to be crashed, recovered alien ships!

What Wood found out from Fisk only increased the mystery and frustration. Ryan sat down with Walter in his den and noticed some books on UFOs and ETs sitting suspiciously amongst the old man's collection. Fisk "had a keen interest in UFOs," recalled Ryan, which is not normally something the average elderly retiree engages in, all must admit. Even fishier, Walter stated he once snapped a photo of an unidentified spacecraft from a window in an office within the U.S. State Department building, in Washington D.C., where he was (supposedly) working at the time. If this remarkable statement is true, *what precisely was his position in or around the upper echelons of government there?* As a news reporter who just happened to snap an impromptu ET quickie? Or in his position as an intel operative, having inside knowledge on an event or UFO flap? Or as something else entirely? Was this UFO picture story a little hint that W. W. Fisk was indeed once an advisor to U.S. presidents, as he had previously alleged, and was much more than just a sticky-fingered villain who purloined an historic ET picture in Kansas and never returned it?

Mr. Fisk then claimed to Ryan Wood that Stanton Friedman got it wrong; Walter wasn't a psychologist, he was a physician, and a surgeon in particular, long ago. Pretty big boasts, with nothing much to back them up with. Fisk startled Wood by accurately remembering Mann's mother as a redhead, calling her the nickname "Red," indicating he had some strong and steadfast memories of that specific period in his lifetime.

William and Floy Huffman moved now and then through the Midwest, from one church or job opportunity to the next; by 1953/'54 Charlette's parents, herself, and her sister were now living in a tiny Kansas town and this may well be where the grandparents ended up as well, at least for a while. Walter Fisk was definitely around in this period, and never denied getting to know the Huffman family fairly well.

Frustratingly, W. W. Fisk had little else to say to investigator Ryan Wood on matters of substance, and particularly on the Huffman family ET photo. He now alleged he never received it or even looked at it! Since he was getting close to ninety years old, could a faulty memory or perhaps some dementia possibly be blamed on the effects of old age? Or was the elderly man deliberately lying? Obviously evasive Walter was not under oath, and Ryan was not a law enforcement officer, so "the whole truth and nothing but the truth" was not necessarily told by Fisk. Flustered, Ryan Wood left after a half-hour chat, no closer to solving the mystery than before.

Wood soon contacted Charlette Huffman Mann at her home in Tyler, Texas, and filled her in on the latest developments. She determinedly set out to both write and call Walter Wayne Fisk to get the honest lowdown directly from him at last. Unfortunately and suspiciously, elderly Fisk refused to reply in any way to Charlette and maintained this sphinxlike attitude towards her for the rest of his days, which were numbered indeed. Gradually slipping into poor health, Fisk was eventually admitted to Albuquerque's Veterans Hospital, where likely only family members and close friends who were totally unaware of the Missouri mystery visited him. If he ever confessed late in life to any involvement in the special alien photo's disappearance, it remains untold and unknown.

Sometime before he passed, as mentioned, intrepid Linda L. Wallace also contacted Walter Fisk for a telephone interview. She received the same exasperating treatment, perhaps even unpleasantly more so. This time, the grumpy, less-than-forthright old vet claimed he did look over the priceless alien picture, but gave it back to Guy Huffman and did not know its whereabouts, thus reverting back (somewhat) to his original story. Defensive Walter spoke highly of the Huffmans and admitted that he had been in Guy's Kansas town retail furniture outlet, but that was nearly all of substance the slippery eel would reveal. But speaking of sea creatures, Walter Fisk did admit to Linda that he brought the alien photograph to the attention of a friend back in the 1950s, "a marine biologist." Fisk claimed that this friend speculated on what the creature might be; perhaps something found in the world's oceans that somehow weirdly (and dubiously) morphed into something else and was then displayed on land by two men. When Linda asked if the biologist was *shown* the snapshot to make this speculative assessment, Walter denied it and suspiciously stated that he only described the image to the marine expert. Then gave the photo back to the Huffmans, which Charlette Mann certainly disputes.

The old MO41 exposure was clearly not something crusty Mr. Fisk dared or cared to discuss. In fact, at some point he didn't want to talk about anything at all. "I'm not saying another word!" he would thunder to polite Linda, clamming up after saying very little to start with. He ended the phone interview with "Don't ever call me again!" An odd reaction for someone if they had behaved with the best of intentions *and* actions back in 1953 and beyond.

While researchers pursued information about Walter W. Fisk, his obituary appeared, via from *The Pratt Tribune* in Pratt, Kansas, where it said Fisk once lived, along with two other small Kansas towns. Pratt was also near where he was buried in early August of 2012. There would be no further interviews or direct information from puzzling Mr. Fisk now. He had died at age ninety on July 26th, 2012, in the New Mexico veteran's hospital, leaving behind many surviving family members, although the obit mentioned Walter lost his wife some time before, plus a brother.

In the aftermath of his demise, it is natural to expect that Walter's personal possessions were inspected by his kin and dispersed as they felt appropriate. Any photographic proof that could have been left behind is likely either now in someone else's hands, or is just plain lost to the world. It seems unlikely he'd have been allowed to keep a copy of the Huffman family ET snapshot; that was intelligence agency property, no doubt. But you never know what secret copy could possibly have been covertly stored by tricky Fisk in shoe box, safety deposit box, or back yard hole.

Two listed obituaries on slippery Fisk stated he served his country by joining the United States Army and seeing action in World War II and the Korean War, but that his main occupation in life was as a photographer for "U.S. News Service." No mention was made of him being an intelligence operative; a trained surgeon; a psychologist; a presidential advisor; a State Department employee; an insurance agent; or even a man who once held history in his hands. His birth was listed as September 8th, 1921, making him nineteen at the time of the actual April '41 crash, just before he served in nearing World War II. Mrs. Mann cannot recall precisely what year this theft happened in, but when she was "around ten or eleven years old" meant that it was likely sometime in 1953 or '54, right when Mr. Fisk was back from his tour of duty in Korea, as his obit stated.

Linda L. Wallace, the researcher from Ohio (born in Sikeston), spoke to a Fisk relative after the elusive, grouchy man's death and managed to learn more about his perplexing career. She learned that he apparently worked as an insurance agent for some time. or was he? Walter was also apparently involved in the shadowy, deceptive world of *military intelligence*. It seems Walter started out in the Navy, and later transferred to the Army, in their intel unit that undertook secretive work Walter could not share in detail with anyone in his family, according to Linda's source. Walter had served in World War II and then the Korean War, this relative asserted, in an intelligence position. Then after a few years of mysterious civilian goings-on, in 1956 he abruptly moved with his wife to southern California, to work for *The Los Angeles Times Mirror*, ostensibly as a newspaper photographer. Yet this Fisk kin, speaking directly and candidly to Linda, told of Walter's wife wondering aloud during this period why he kept conducting some sort of unspecified business at a nearby L.A. army base. Ms. Wallace discovered Fisk's spouse at the time had declared that she felt sure her husband was *still* an Army spy, an undercover operative who was actively involved in secret projects, despite the fact that the most previous war was over by *years* and he was *supposed* to be out photographing general news events in the greater Los Angeles area.

So imagine that, Walter Fisk was apparently in reality a busy peacetime "spook," a well-trained slippery fox, spying on his fellow Americans for the Army higher-ups, at least within their intelligence circles. And that may well go for his work in small town Kansas as a seemingly unsophisticated and earnest insurance agent in the early 1950s, and then digging into news events in southern California in the latter half of that decade.

In the 1960's Walter Fisk transferred to Washington D.C., where he worked for United Press International as a photojournalist, his beat mostly at - of all places - the White House. Now he was suddenly trusted around the most famous and powerful men in the world! This would have been during the liberal Kennedy/Johnson years. JFK was in the Navy as an intelligence officer in 1941, around the same time Walter Fisk joined. {For Fisk to have abruptly been transferred to Army Intelligence is a fairly clear signal he had been involved in Naval Intelligence first, and that he possessed the proven ability to cloak himself in an outer layer of respectable innocence.} Now in the mid-1960s Presidents Kennedy and Johnson were the most powerful men on the face of the earth, having access to much inside military intelligence information and perhaps the U.S. space program's juiciest secrets. And some days, none other than Walter Wayne Fisk was around, sniffing around for news, camera in hand, quite possibly *still* reporting back to his military intel superiors. Yes, the plot had thickened considerably for Fisk, the mere small town insurance peddler just two decades before.

It was Walter Fisk's old newspaper, of all ironic sources, that may well have outed the modus operandi of the Army intel plan - involving agent Fisk? - to trick some Americans into revealing but private important matters. In September of 2000, *The Los Angeles Times* ran a story that described opening up declassified military files to expose a special top-secret "Insurance Intelligence Unit" that operated in the 1940s and beyond. As displayed by Linda Wallace in her fact-collecting e-book and web site on the Missouri incident (and other UFO claims), this ingenious Army intelligence program was run by General William "Wild Bill" Donovan and his business partner. Bear in mind that Donovan - appointed in mid-1941 FDR's "Coordinator of Information," he knew so much - *was specifically named in print by the president as someone with inside knowledge of the "new wonders" of recovered "celestial devices" the U.S. military covertly held,* this within a top secret memorandum from President Roosevelt in February of 1942 for Army Chief of Staff George Marshall (see Chapter Six).

Donovan's ally in the insurance project was American International Group's insurance mogul, Cornelius Starr. The two powerful men used insurance agents directly or indirectly to learn more about pre-war preparations and then wartime operations in Europe to help the allied nations win that complex conflict . and then they kept their secret program going to spy on *Americans* and others in the years afterwards. The *L.A. Times* article said that Donovan's post-war intelligence unit and the AIG headquarters even worked out of the same office Los Angeles building. *They were mining data from the field of insurance forms and reports* in order to shape intel operations, supposedly for the good of the country. Could this L.A. headquarters and undercover operation be what Walter Fisk was involved in, back in the 1950s? The enigmatic Fisk supposedly didn't move full-time to the City of Angels until '56, but he could have been moved around before and during that time on covert assignments like a pawn on a chessboard, working anywhere in the country (like Kansas) on whatever procedures that were deemed necessary by his L.A.-based intelligence superiors.

In the 1940s, America's hush-hush, highly classified nuclear program created and refined in New Mexico, at Los

Alamos and Sandia science laboratories. It was an unqualified success, resulting in the "super weapon of war" that was dropped on Hiroshima and Nagasaki, Japan, to destroy those cities and bring the Imperial Japanese forces to their knees. The Manhattan Project didn't win the war single-handedly, but it certainly helped halt it in the Pacific theater. But we shall see in documents later a few statements that described what crashed in Missouri as apparently utilized by top American scientists in that same Manhattan Project, as it was code-named. It seems like a pretty a wild allegation, but if it is true, it further explains why Show-Me State alien technology captured was eventually flown to New Mexico for top secret study. Some of it may never had made it much further than the Kirtland/Sandia/Manzano area, where once again Walter Fisk and his immediate family dwelt just a short drive away, towards the final decades of his life.

As the very end of the 1990s approached, Charlette Huffman Mann agreed to travel with researchers Stan Friedman and Ryan S. Wood to Washington D.C., to personally search for more evidence. The two men pored over some yellowing records in the National Archives but found precious little in the short time they were allowed to poke around, and nothing really regarding any detail of hushed MO41. However, they *did* manage otherwise to get their hands on a very telling, even downright shocking once-secret report from 1947. It was a classified document not from the archives that discussed some enigmatic, foreign materials recently collected from the New Mexico desert outside Roswell, New Mexico, that steamy July, and what they meant to government and military investigators. Charlette, Stanton, and Ryan inspected the fascinating, recently-leaked "White Hot Intelligence Estimate," which was originally typed up by key Air Force personnel, under the direction of its author, General Nathan R. Twining, head of the "AMC" or "Air Material Command" and the

"Foreign Technology Division" based out of Ohio's Wright Field. This eye-popping report was evidently sent to the White House, in late September of 1947. The nineteen-page document was obviously intended initially for just a *very* select few in government to see. In fact, it was originally not for a single soul to lay eyes upon unless he first had the "express permission of the president," Missourian Harry S Truman, who was its critical original recipient. Other previous, later-leaked documents state that Truman expressly sent Twining to New Mexico to reinspect the newly-recovered alien artifacts there, to give the president an accurate depiction what was being found, analyzed, and decided upon by knowledgeable military and scientific personnel concerned. What was learned undoubtedly continued to reverberate in clandestine U.S. intelligence circles in the early 1950s, emanating from high-tech New Mexico laboratories outwards.

What was printed within the resulting Twining "White Hot" document's astonishing pages were direct references that back up once more the notion that the 1941 Show-Me State crash material was a) extraterrestrial; b) used in sensitive nuclear weapon and atomic energy research in New Mexico; and c) also utilized in building a special new experimental American aircraft. It was in short a riveting blockbuster report worthy of great dissemination and discussion, then and now.

Completed in great secrecy on September 19th, 1947, and signed by a set of powerful military leaders (and readers) five days later, Twining's super-sensitive "Intelligence Estimate" was officially titled *"Mission Assessment of Recovered Lenticular Aerodynes."* The numbing account mentioned the many UFO sightings by various pilots earlier in 1947, then a description by General Twining and his top aides of the materials culled from the recent Roswell regional crash in New Mexico. "Aerodyne" was an early term - very old-fashioned by today's terminology - for a spaceship. In this case, the aerodynes concerned were only ones the U.S. Army had somehow "recovered," in secret operations, thus resulting in the "Above Top Secret" label for the classified report.

"Based on all available evidence collected from recovered exhibits currently under study," General Twining dictated to his security-cleared secretary, then went on to list eleven high-level organizations, including the U.S. Army Air Force, the RAND enterprise (partly headed by Hap Arnold - see Chapter Five), the Atomic Energy Commission, and the Massachusetts Institute of Technology (Van Bush's sometime university), among others. The army-recovered 1947 New Mexico hardware was "deemed extraterrestrial in nature," he continued in another sentence, a stunning declaration in itself. How could those near the top of the armed forces putting together the report be so certain it was from another world? The answer was based on our specific Midwestern subject matter.

"This conclusion was reached as a result of comparisons of artifacts" .leading to a brief *blacked-out section* where the words *"from the Missouri"* fits in perfectly. The crucial sentence finishes with the heart-stopping phrase "discovery of 1941." Bingo! *A comparison of artifacts from the Missouri discovery of 1941*, imagine what that could be! While the phrase is a "smoking gun" in this case, alas, with the redaction it is unfortunately reduced to a *fairly* smoking-hot clue. What *else* besides MO41 would have been "discovered in 1941" that would have been "extraterrestrial in nature"?

{Some of the same questions should be raised when pondering what a German-born scientist named Otto Krause allegedly asserted when he stated in 1962 he once learned of a special classified "propulsion device," according to two different contemporary researchers. This remarkable item was created from the atomic know-how within aerodynamic alien technology recovered from *"an air crash in Missouri in 1941."* Such statements assuredly *have* to be direct references to what fell to earth just outside of quiet Cape Girardeau. Physicist Krause supposedly added that to his knowledge this special navigational device that was created from copying the ET craft components was "more top secret than the atom bomb itself." Simply *stunning*, if true.}

Yes, Charlette Mann and her two investigative friends were riveted when their eyes rested on the Twining report's key "1941" sentence. *Pay dirt!* It's important to stop and ponder another critical point here, regarding the report's description written for the president. *Nothing in the document was elaborated upon regarding "the discovery of 1941."* That is, the decorated Army Air Force general who authorized and guided the report did not feel the need to take the time to explain to President Truman or his top aides and leading military leaders who signed on to the report precisely *what* the "extraterrestrial in nature" property from '41 was. *Obviously because Harry Truman already knew, and likely so did his eminent advisors.* It was simply a matter of fact that by '47 - and probably some years before - that Mr. Truman was very familiar with MO41, and would quickly recognize the two simple references to it within the top secret document. Harry and his most trusted consultants needed no special Twining briefing on the '41 discovery, nor further paragraph explanation or exploration. It is also clear that General Twining was familiar with what President Truman fully understood on the MO41 affair, and probably also the small band of named others qualified to help write and also read his report. To them, the exceptional Missouri UFO crash was merely common - but still secret - knowledge.

Summing up this amazing declaration under the section entitled *"Part III: Scientific Probabilities,"* Twining notes "The technology is outside the scope of U.S. science, even that of German rocket and aircraft development." This eye-catching statement, along with a second reference later on within the highly classified document, convinced Charlette Mann around the turn of the new century/millennium that her grandfather certainly *had* been telling the absolute truth all along. Advanced aliens really *did* crash near Cape Girardeau in '41!

In returning to the New Mexico connection, one cannot underestimate the importance of "AEC" listed in the 1947 classified Twining report; the Atomic Energy Commission reference certainly reinforces the notion that "atomic secrets" were found within the power unit or propulsion drive engine of the extraterrestrial craft found outside Roswell. Then this material was placed "in comparison of artifacts" from MO41, bringing us back to the startling possibility that our nuclear bomb development project of the 1940s was being aided by the secrets unwound from within recovered alien technology. More specifically the power source components found within the engine, backed up by the aforementioned respected German physicist Krause's remarks some years later.

Another Twining report paragraph is again marred by redaction, but starts out referring to the "mode of operation" and the "instrumentation" within the recovered Roswell-area craft that was comparable to something else that was being tinkered with by scientists in the mid-1940s: *"the aerodyne from reconstruction of available wreckage."* Then. more blacked-out statements cloud and frustrate the reader. The completion from the redaction indicates the Roswell-recovered alien ship's engine system ran on "biosensory and optical stimuli." In other words, the '47 vehicle was commanded by ET pilot thoughts and visual interaction with an extremely advanced mechanical power and control system not fully understood in the 1940s, and perhaps only partially fathomed today, but still better understood when coming back to our academic comparisons to what MO41 intelligence findings yielded. Or to speculate reasonably: the two UFOs were similar technologies, perhaps from the same race of interstellar explorers. And what is more, they had similar crash circumstances and causes, debris fields, craft damage, and pilot deaths.

The specific redacted Twining report "aerodyne reconstruction" remark certainly seems like yet another reference to MO41; what other "available wreckage" would be utilized in testing obvious alien hardware by 1947? Perhaps some of the alleged Los Angeles area recovery from 1942, but that's about it. We must bear in mind here through another leaked document - this time from a retired CIC/CIA agent named Tom Cantwheel - that the United States Army Air Corps was secretly trying to build a round, replicated spaceship made from the parts of *"an aerodyne recovered in 1941 that crashed in southwestern Missouri,"* and supposedly from a second otherworldly craft found the following year. {Yes, the term "south*western*" was a mild mistake; Cape Girardeau is decidedly "south*eastern* Missouri.} This UFO wreckage reconstruction and secret new U.S. vehicular flight testing allegedly took place in 1945, under the direction of notorious General Curtis LeMay, the 1990s confessor Cantwheel told us elsewhere in his typed letter... so the timing fits with the "White Hot" statements.

Further into Twining's "Intelligence Estimate," on page sixteen, under the heading *"Part V: National Security Structure,"* the general dictates the following riveting, blockbuster statement: "Even the recovery case of 1941 did not create a unified intelligence effort to exploit possible technological gains with the exception of the Manhattan Project." *Wow!* Eye-opening confirmation yet again of Cape Girardeau's recovered craft and its very significant impact on the American atomic bomb! Another smoldering hot pistol, very nearly a smoking gun again. General Twining is essentially complaining here to President Truman and his top team that the MO41 materials were never fully enmeshed into the world of covert government intelligence and military strategies, along with academic, corporate, and scientific opportunism, not like the ongoing operations for comprehending the July '47 New Mexico materials that were "currently under study" by the whopping eleven organizations mentioned earlier. But once again, "*the recovery case of 1941.*" What a stunning tell-tale mention! "White Hot" indeed.

Charlette Mann read the Twining report once intended only for President Truman and his top advisors and pondered its shocking meaning; she was firm in her conclusion, especially after learning of the serious testing of the original assessment's makeup (both the wording in print and in style, and the very paper it was typed up on). This important aspect was painstakingly researched by UFO document experts Stan Friedman, Ryan Wood, and his father, Dr. Robert Wood, PhD. That esteemed investigative trio ended up also feeling strongly that the "White Hot Report" is completely authentic and worthy of very serious discussion - and acceptance. Charlette agreed. "I have not forgotten holding that paper in my hands and realizing that my family's story was real, was solid, and for me was an answer to a long time question," she summed up with great relief and pride. Indeed, it seems certain now that Twining's amazing secret assessment is pretty solid printed proof that the U.S. Army knew all about - and controlled for a while - the Missouri materials that Reverend William Huffman first spoke of in April of 1941, and his granddaughter first spoke up about on a greater public scale some fifty years later.

In 1982, General Twining passed away at age 84. Before he left this world, he is reputed by some sources to have confessed to his son that the U.S. Army did indeed retrieve alien cadavers from crashed space vehicles, although allegedly the main focus of his alleged near-death statements was based upon Roswell. And to add to the murkiness and untrustworthiness of these supposed admissions, Twining may have been suffering from possible dementia at the time. Even worse, Nathan's son, Nate, gave conflicting answers as to what his dad spoke of privately in his final months alive, but if MO41 was mentioned, we may simply never find out the real truth what the fading ex-Air Force general learned or talked about. The exceptional MO41 story just wasn't well known by even rudimentary UFO investigators in the early 1980s and was perhaps ignored by Twining caretakers as mindless blather, if uttered at all.

Overall, however, those who have looked into the MO41 crash claim now feel more comfortable at last about truly believing what the granddaughter saw. And what a dutiful Cape Girardeau photographer (G. D. Fronabarger?) once snapped and saved, then handed to a trusted pastor. And quite possibly what undercover operative keen on photography may have deliberately stolen to bring to his intelligence superiors. And what still tantalizes interested researchers today, sleuthing with an open mind for just the right "ground zero" outside Cape Girardeau, searching for remaining clues to the alien visitors. Could contemporary answers to pinning down the precise MO41 crash site simply be found with just a few clicks on the modern World Wide Web?

CHAPTER THREE

Startling *Topix* Clues

"My grandparents owned and farmed the land where the crash occurred."

In the late winter of 2012 and the early spring of 2013, the stunning subject of MO41 sprouted in the fertile online soil of *"Topix,"* a national computer website that features conversation posting boards, extending to each American state's localized areas. Exciting-sounding new ET crash clues popped up in web forums regarding life in Cape Girardeau and also in neighboring Scott County, Missouri, with some enticing, fresh information and opinions bubbling up to the once-frozen surface. From cyberspace came entrancing traces from outer space. Not all of these *Topix* postings can be taken as gospel and valid, of course, but many seem highly intriguing and worth further investigation. Within two different post boards and a few separate public threads, two main enigmatic sources provided the most riveting allegations and clues yet, perhaps giving us the biggest breaks in the case since Charlette Mann's initial foray into revealing in the 1990s what her grandparents knew and discussed.

After several people posted some initial questions, general comments, and innocent theories on MO41 in the *Topix* message sections for Cape Girardeau and nearby Scott County, in came some better-informed, more serious posters, seemingly helpful to a degree in educating others on just what they knew to be true in the astonishing case.

Almost all of them, however, agreed that as they understood it the actual 1941 alien accident site was located just *south* of Cape Girardeau, both the city and the county, down into neighboring Scott County. This would have been rural farmland about located thirteen to fifteen miles south of William Huffman's North Main Street home in downtown Cape, thus neatly fitting the known criteria for the locale. If the patient pastor and his police pal were driving at night, taking rambling roads and twisting turnoffs as described, and were headed for this obscure northern Scott County countryside locale, it would have taken about thirty or more minutes in the dark of night without encountering any memorable sites or landmarks along the way, which is precisely how the Huffman narrative recalls William's unusual odyssey (see Chapter One).

In 1941, only a few paved streets with mostly farm-related businesses were in place in these unimpressive, quiet Scott County burgs, and on a Saturday night almost all of them would have been nearly empty or closed, with citizens either home eating dinner and relaxing. Quite likely many small town folks were out and about in nearby Cape Girardeau, where there was far more to see and do. According to period photos, nearly all buildings in these hamlets - like Chaffee - were one or two stories, surrounded by undeveloped lots, hemmed in by forested areas or miles and miles of crops. Gravel and/or dirt roads dotted the landscape outside towns. There was very little police force or a firefighting department in these villages, counting instead on volunteers in case of emergencies or perhaps the use of the better equipped Cape Girardeau city or county services not far away. Thus if something did crash from the sky out in this particular bucolic territory, it would have been perfectly natural for the property owners and/or concerned citizens to have picked up their telephones and dialed for the Cape Girardeau police and/or fire department, or at least for a Cape switchboard operator to connect them to that site. So another criteria for the tale has been met by clues that lead us in the direction of "down in Scott County," even if just barely.

In Scott County, to the east and not far from Chaffee, Benton was the county seat, where Sheriff John Hobbs and his deputies moved in and out of the courthouse, jail, and justice center. Benton featured just a few thousand at most residents, many of them spread out on isolated, rural farm properties without a lot of close neighbors. Cotton was king in Scott County, or nearly so, but corn, soy, and other crops grew tall and full around much of the area.

"I asked my mom about it," a 2010 poster wrote on a Scott County Topix site, but the parent evidently didn't know a thing - but one of her friends *did*. The 1941 extraterrestrial calamity claim was *real*, the pal explained, but "it was covered up because WWII was gearing up, and the government didn't want anything drawing attention away from the war effort." Sometime later another tantalizing *Topix* post came in, claiming that the UFO crash site "is on a friend of mine's property. There is no problem looking at it if that is all you do." Seemingly an electrifying statement and offer, but sadly the online source never returned on the public forum to either disclaim this post or back it up with further explanatory information or invitations.

And before we continue, we must always remember: *anyone can post anything on the internet, true or not.* Still, we're just getting warmed up for what juicy nuggets arrived next.

Another highly intriguing source who seemed to be armed with genuine, detailed knowledge on MO41 posted on *Topix* in early January of 2013. Dubbed "Just Wondering" from "Cleveland, Ohio," this mysterious unnamed person wrote back to those on the message board who deemed the suggested Chaffee rural community as the site of the alien spaceship's impact: "You're getting warm!" J.W. (as we'll call the source) elaborated at another time with: "Yes, it was in the Chaffee area, and there is a man from Chaffee that has a piece of the wreckage. It was always referred to as "the Cape Girardeau crash" to throw off investigation." Well, well! Keeping in mind the possibility of fraud, this certainly *sounds* like someone with the inside scoop, coyly cluing us into the once-secret alien crash locale!

If this blockbuster online statement is valid then not only is there a possible descendant of the incident who has clung to priceless souvenirs. but he - and possibly others - apparently doesn't want to be found by outsiders and makes deliberate attempts to mislead others who attempt to sniff out the story. He evidently doesn't want his name bandied about in the media or have his apparent recoveries placed in the public arena for scrutiny, for whatever reasons. Overall, the J.W. declaration about this local man and the MO41 incident was riveting. Of course, anyone can say anything about any topic on the internet, and there's no "truth detector," but "Just Wondering" sure kept attention from wandering.

"In the Chaffee area." This is still rather broad, but if true narrows down the search considerably from the initial vague Huffman directionless recollection of "ten to fifteen miles outside of Cape." But was the crash site south or north of Chaffee? East or west? J.W.'s statement at least tells us that the cosmic affair likely did not happen *in* Chaffee, the more populated village, but was not too far away so that the town could be used as an identifiable marker, or somewhat of a starting point today. But it also tells us semi-helpful J.W. was not in favor of just coming right out and telling us the all the facts in a clear precise manner. Such as a proper farm address, nor the name of the local gentleman who clung to his otherworldly prize

possession, and where he resided. Enticing hints were all J.W. was offering.

The online guessing game continued with yet other comments, led by another source claiming "I'd always heard it was near Chaffee and Benton." Another stated he learned "it was between Chaffee and Scott City." The general forum consensus placed the site as *at the least* between the hills east of Chaffee and old Highway 61, narrowing the field of possibilities a bit.

The enigmatic online Ohioan further titillated with other *Topix* remarks in early 2013; J.W. claimed that there was a specific sign of sorts created on or near the area of the crash site. A potentially helpful clue, to say the least, but precisely what did this special marker consist of? Where precisely was it located? Maddeningly, "Just Wondering" wouldn't reveal details. "Where ever something tragic occurs a memorial is normally placed, the generation after those fine men that witnessed that tragic night, and yes it was very tragic for them, held that area very special. The generation now just sees it as something that's always been there. Be respectful of what you find," the poster somberly advised. J.W. seemed to believe that anyone who goes looking for the old alien accident scene *will* find this unknown memorializing indicator still standing. Interesting, to say the least! Frustratingly, there was no specific location given by J.W. for this alleged "memorial," no address or farm road listed, and no description of what it looks like.

To explore this thought further, "Just Wondering" strongly indicated that there was once a group of local gentlemen (and women?) in or near Chaffee who were deeply affected by the extraterrestrial crash, feeling it "tragic." These were perhaps either first-hand eyewitnesses, or heard about it from someone who knew of the crash, and were deeply moved by the sudden loss of life, even if it was extraterrestrial and strange to them. And they were reinforced by the next generation after these first responders, likely meaning the offspring of those who went to the crash site, taking action twenty to thirty years after the fact. This alleged exclusive club of sorts actually *did* something afterwards about their lingering memories and feelings.

Since rural Missouri was - and remains - a very Christian community, we can therefore logically conclude that a "memorial" to those who involved would be a cross. Or a headstone with perhaps a cross (or three) or other Christian designs, or even masonic symbology. But is this supposed special object a salute to the men from the '41 site - who had since died - or to the ETs who died right in front of them? Or both? By decades later, this memorial would naturally be faded and weathered, following intensely hot summers and cold winters, driving winds and rain. If the memorial is still in place, it must have been made of a hardy material, likely stone or cement. Metal would have rusted away by now, and wood fallen into rot and disintegration.

So it is very conceivable that the original MO41 crash locale response crew began dying off as the decades melted away and their children who knew the story got together with a few hardy survivors and quietly accomplished something substantial about the momentous event. They marked its anniversary perhaps, or elegantly immortalized it all in some subtle way without the United States government - and the intimidating Army - knowing about it. And since then "the generation now just sees it as something that's always been there." The memorial means nothing to them as it was never explained as such; it was evidently just something they'd see while driving or walking past in or near Chaffee and thought nothing much of it.

The original 1941 crash collective "in the Chaffee area" didn't let their great secret out in book or public record form. so one must ask *exactly how did "Just Wondering" find out about all this?* How did he know the 1940s men so well that he could describe them being "fine" and noble in their attitude towards the alien deaths at the "very special" crash site? How did J.W. know about the "next generation" taking action afterwards? How could he emotionally ask for continuing respect for the accident location and memorial tribute if he *didn't* know all about it from first-hand experience? Why did he obviously monitor the *Topix* board for the "Scott County UFO Crash" and respond somewhat enigmatically to only a few intelligent postings after he apparently helped keep the subject under wraps himself for so long? And then why did he clam up afterwards?

Despite prompting by a few posters, the mysterious, anonymous "Just Wondering" stubbornly would not come forward with any more clues or comments, indicating he may well now feel as if he has said too much already. If the memorial marker claim bears out and the crash site is found in Scott County by intrepid researchers, the cryptic J.W. may well come to regret his stated private opinions on the public message board, at least if the marker and the farm site are not treated well by single-minded treasure hunters and vociferous fame seekers. One of J.W.'s final posts was to suggest that the matter might be better left alone, then his online *Topix* trail went cold.

Who exactly would have made up the group of local "next-generation" Chaffee people who concocted this memorial? If it contained Christian symbology then God-fearing, church-going farmers and their families come to mind, perhaps even whomever owned the land for the MO41 event. And possibly their close friends and/or farmhands. Maybe a few others who were bossed around by the Army at the site that night, such as 1941 area firemen and policemen, and their offspring. With a cross or a religious memorial, they might have been itching to gain a kind of quiet "spiritual revenge" on the government for ordering them off the property and into general silence. The entire plot to create something to honor those associated with the crash - even if it was just simple generic wood, stone, or metal artwork - might have been a cathartic, helpful bonding experience for the next-gen group. They felt so strongly about MO41 they had to have spent some amount of time, money, and energy on the project, which might have needed community approval before its installation, whatever it is. It's frustrating not to receive more information from "Just Wondering" to get the entire story on this angle of the remarkable saga, but at least we can speculate reasonably to gain a greater understanding of the situation. {Interestingly, a generation after 1941 (built in 1976) a large white cross was erected by the city's Chamber of Commerce on a hill overlooking Chaffee and remains there to this day, surrounded by thick trees, but that is certainly no proof it is in any way related to MO41.}

Another helpful person on *Topix* stepped forward during the ongoing internet discussion. This different, authoritative poster named "AllSouls" showed himself to be a most knowledgeable forum poster with eye-catching remarks on MO41. This mysterious online source began issuing fascinating statements in late March of 2013, evidently for the first time. It was obvious from his comments that the region's UFO crash was a topic that really meant something and he had personally investigated it. Many were eagerly asking: "*Where* did the '41 accident take place?" "AllSouls" felt he had the answer. "It was four miles west of Grammar," he wrote {substitute fictional town name used}. This would place the crash not too far from the Cape Girardeau airport, but near where "Just

Wondering" hinted was the correct locale: "in the Chaffee area."

A new bombshell was then introduced by the *Topix* poster named "AllSouls." He submitted: "Portions of Crash material was possibly stored in barn torn down 5 yrs ago." We'll leave aside criticism of the errors in sentence structure, spelling, and capitalizing, and move on to the subject of this startling, assertive post. His declaration indicates strongly that "A.S." (as we'll call "him") had been investigating in person. He'd been talking to locals and collecting data, and finding out juicy facts, yet this poster with no real name given listed his writing location as "Owensville, MO," which is a small town over a hundred miles away, to the west and north of Cape Girardeau. Was this the truth at last, from an actual Missourian in the know? Had A.S. recently traveled downstate to the crash site community? Where he learned that additional wreckage was once found and tucked away *after* the initial accident recovery by the military? If this is true, it's a tremendous story, just waiting to be uncovered and explored for international consumption.

Certainly the eyebrow-raising remarks expressed online could have been purely a hoax, a trickster having fun at the expense of people intrigued by UFOs, always targets for the cynical and opportunistic soul. But if that were the case, wouldn't such a devilish person create much more flowery, well-designed, outlandish and/or pompous assertion? Wouldn't his statements have been more detailed, better-expressed, lengthier, less obscure, and more creative than *this*? A.S. was claiming he had learned actual alien debris was at some point hustled into a barn - "possibly." An odd equivocation. Another unusual detail: that the barn was then torn down just "five years ago," which also shows us how recently A.S. had been in the area, doing his research. Very recently. It is obvious

MO41 gnawed at him, and tantalized him, this proof of life beyond earth, visiting here.

"AllSouls" provocatively added in the same internet posting: "Owner won't discuss what was in barn when he bought property. Lives in area." Another solid indication of some personal research by someone in the know, interacting *with* someone in the know. But why and how could someone have all of this inside scoop? Exactly how could this assertive source have known just where to look and who to talk to? And what was the farm owner's name he spoke to?

A few answers seemingly arrived on April 3, 2013. A.S. was back, and at first writing an apparent correction to his previous *Topix* statement: "I posted earlier an error as to location as I understand it. I wrote 4 miles west of Grammar. I meant to write EAST. Sorry for the confusion." *Well what the deuce?* This abruptly altered positioning leads us to a "south of Cape" location that does not really fit in with many other recent descriptions. Such an eastern county locale would be close to the Mississippi River. Was this new, adjusted site proposed by A.S. accurate? Or was he in reality purposely misdirecting people away from the true locale, perhaps for fear it would be discovered and exploited by less-than-scrupulous UFO buffs? Either way, A.S. then went on to knowledgeably discuss the Parks Air College, based out of East St. Louis and in Sikeston, which indicates to us again his continued advanced insight and research into the overall crash storyline. Those two 1940s flight training schools were evidently involved in the recovery of the materials (see Chapter Five). A.S. also mentioned his awareness of vintage airplanes used and stored at the Cape Girardeau Regional Airport, reinforcing the notion that "AllSouls" did plenty of physical ground research in the region. It may also be another clue that the crash farm was not far from the airport, which is precisely what "four miles west of Grammar" tells us. At any rate, it was time for A.S. to drop more direct clues to the great event.

"Anyway the crash site in 1941 was immediately sealed by military and debris was cleared from the area." This was no great revelation; it was public information that had been on the internet and laced within a few books over the past decade or so. "After war remnants were discovered when cistern dug close to site." So it is clear once more that A.S. had been speaking with someone pretty close to the case, if not directly involved. Was this the very same local man that "Just Wondering" was describing? The gent who supposedly possessed "a piece of the wreckage" and purposely threw people off of the trail by referring to it as "the Cape crash"?

"*After war*" can reasonably be described as the period following World War II, in the late 1940s and '50s. A "*cistern*" is an underground collection tank for rainwater, most often when absorbing the runoff from a barn or house. A drain pan, of sorts, helps above ground, to gather falling rain in the cool tank that can rest below ground, or simply dug in at its base. The rainwater helps later, to be drawn up like water from a well, to be used to water crops and houseplants, or for livestock. Or for cooking, washing, or even fighting fires. It was pretty essential to a farm without good plumbing or a sound well. One thing to keep in mind from a cistern, however, is that if this is a valid MO41 clue, it strongly indicates that the fiery crash took place in a farm property's *yard*, not really out in a field. It seems highly unlikely a cistern would be dug into a crop, which must be tilled, planted, cultivated, and picked. A cistern would be in the way there, but not placed in a back or side yard, likely pretty close to a structure that could drain its rainwater into it.

"Remnants" indicates that the debris removed from the ground during the post-war cistern tank dig were undoubtedly unusual metal fragments that remained from the original spaceship crash, not just any metallic junk or rotting old soda cans found buried in the farm's soil, of course. The ship's impact with the ground that April '41 evening apparently broke it apart and drove several strange shards down into the soil, out of reach by a surface recovery effort by the army.

Physical remains, still imbedded or buried at the site. Maybe a few small shards from the ship's exterior, or a broken section of its inner design or support chassis. If such speculative daydreaming is true, this is one exciting revelation! And here's a further thought: *that there continues to remain undiscovered even more extraterrestrial prizes within that patch of Missouri soil,* just waiting to be dug up and properly tested. Spacecraft technology beyond mere shards, such as genuine alien tools or gauges, weapons or clothing, perhaps even the corpse of a fourth member of the broken ship's crew. Shocking physical, tangible proof to show the world that aliens are quite amongst us, or were in 1941. The notion is astounding, exciting, and perhaps quite possible.

All of A.S. revelations leads one to believe they mesh pretty reasonably with what J.W. was revealing. Was there a special memorial near the cistern site, or at least somewhere on the property that was sold to the new owner without revealing the marker's purpose? Certainly the alien crash was not explained to the new owner. If the new owner simply allowed the memorial to stand, quite unwitting, could this mean there remains *two* great signs or indicators left nowadays for researchers to find in farm country? An old cistern (perhaps rusty and decrepit now) and the aging memorial. if such helpful clues were found together on a property in the described region, they might well lead an intrepid explorer to the proverbial Promised Land.

Contributor "AllSouls" was getting amped up. He continued on the *Topix* forum, stating the odd materials from the alien accident, found some years later, were "Stored in barn as new owner didn't know history of the crash." Could the knowledgeable poster's source for this information have been the man who purchased the farm property without being informed about the alien encounter and army retrieval some years earlier? Or was it someone who simply knew that particular agribusiness owner/operator well? What was evident from this statement is that the previous - and knowledgeable - owners of the UFO crash property just wanted out, and understandably didn't want to blow the sale, so to speak, by spooking their interested buyer with weird tales of little gray men and their silver flying machine landing tragically on their domain. And once again, from this declaration we see that the online A.S. *knew* that the unnamed farmer involved took the unusual pieces he had unearthed over to a barn on his property, which is where he likely and most naturally stored most everything he found or grew of value on his fertile land.

"Subsequent property owner tore down barn 5? yrs ago." This next statement reinforced the A.S. remark from days before, indicating his source noted a deteriorating structure and got rid of it, but first emptied the contents, naturally, perhaps safely placing them in a new barn on the property, or in another secure location. Whether the collected crash debris was hidden in any new or other old barn, or was even thrown out, or simply given away, is not clear as the unnamed Scott County contact for A.S. wouldn't give any further location details, perhaps realizing the debris' value and fearing for its safety - and his own. "Will not discuss remnant materials other than resembled aluminum but could be bent and reshaped without creasing," which sounds eerily like what we might call "memory metal" in this day and age. Obviously the old farmer did talk at least a little bit about the contents of his unusual harvest, since it was so stunningly unique. What other metal from the mid-twentieth century could be "bent and reshaped without creasing"?

The amazing A.S. description of the wreckage also brings to mind what some witnesses said the metallic debris from the 1947 Roswell crash resembled and reacted like, unable to be fully bent or folded without snapping right back into place, amazingly. Silvery-gray and seemingly indestructible, unable to be cut but like thin aluminum. And we know further from this A.S. remark that the bemused old farmer who found it had been trying to manipulate the metal, undoubtedly with great quizzical intensity, finding it dazzlingly unlike anything he'd ever encountered before. Without knowing about the alien disc crash, at least at first, he was probably buffaloed and perhaps turned to others on the farm or in the nearest community for assistance in understanding just what he had oddly unearthed and unsuccessfully tried to contort in his rough hands.

But this wasn't all for the remarkable description of the seemingly magical metal from the Scott County soil. It featured "stamp marks on some pieces," strange imprints or designs that were "not identifiable numbers or letters," according to "AllSouls." Here we have the highly intriguing notion of a truly alien language and/or symbology, struck somehow into spacecraft hardware, a kind of hieroglyphic imprinting that would have been incomprehensible to anyone on earth, let alone a busy 1940s Midwestern farmer. And this small but significant detail also fits in with what Reverend William Huffman said he saw when he peered into the cracked-open spaceship: odd symbols and markings on a silvery-gray metallic band that he could not read or comprehend. They certainly weren't English letters or symbols, or any other language either man had ever seen, quite evidently.

"Very lightweight and thin" was how A.S. summed up the described alien hardware recovered by his arcane Scott County contact, "but could not be cut with tin snips." Now there's a startling and yet quaint, realistic reference to an old farm implement: metal cutting shears, which a farmer would have referred to with the rustic term "tin snips." Such a device would have been used on a farm for creating barbed-wire fences or metal holding cases for harvested food or even water (like a cistern). But even these powerful cutters were not able to break the unique ET ship rubble. It was evidently very much technology ahead of its time, even today. The farmer and his family and coworkers (if shown the materials) had to have been completely dumbfounded, and likely started asking questions. It might have been only natural to have brought the amazing items to the attention of others to get answers, yet at some point they were kept under wraps in general, it seems, perhaps when the answer was "it's from outer space, and it's a secret."

"He won't say what he did with the material he found. He's old. Maybe nuts who knows." With that, A.S. wrapped up his eye-catching claims for that day's *Topix* posts, one obviously dear to his heart. He was a MO41 researcher and forum visitor unable to keep himself from revealing some amazing things he'd recently learned. A.S. and his statements were of course a bolt of lightning through that particular chat board and in this historic case in general. Fantastic, exciting new details and researched history seemingly to add to MO41's legend, but was it real? People began writing in, urging more specific information, and quite naturally asking for the exact location of this farm and its owner so that they too could investigate and learn more in person. "AllSouls" - just like "Just Wondering" - did not respond to such further specific site requests, at least not within the confines of the chat forum. Only time, and careful ground research, will be help to nurture and bear the fruit of the planted A.S. seeds, it seems.

Just who was this fascinating poster, "AllSouls"? He described himself on *Topix* as writing from New York City and still other locals in other postings, not just from Owensville, Missouri, indicating he traveled around. Or was he being creative or purposely deceptive in relating his background information, or even in his UFO crash recounting? Did A.S. *really* learn all of this first-hand, and just exactly how? In a posting for April 5th, 2013, A.S. merely displayed knowledge of the Cape Girardeau airport again and signed off, giving no further crash clues. Since "AllSouls" posted on a "Cape UFO Crash" message board for *Topix,* and "Just Wondering" wrote information on a "Scott City UFO Crash" message board, on different days and times, could they have been the same person, hiding in anonymity to purposely cloak his identity? J.W. did manage to pop up long enough to reply to this very question, simply saying no, they were *not* one in the same.

On April 7th, 2013, "AllSouls" was back, with other opinions and a further jolt of inside lowdown on the MO41 incident: "This much I do know because my grandparents owned and farmed the land where the crash site occurred.*"* Wow! Now we have the real reason why A.S. was so fascinated! It was personal. *His own blood relatives had legal and emotional ties to the land, supposedly, but sold it perhaps even before A.S. was born.* He had heard of the strange saga at times and pondered its implications and maybe resolved to do something about it someday. This is why the message board source took the time and money to do so much researching and digging, perhaps literally, and how A.S. knew where to look for interviews and information. Supposedly his own kin experienced MO41 firsthand, then later spoke of it in occasional bits and pieces, and piqued their grandchild's curiosity for life. Now he was ready to share what he discovered with the general public, or at least those open-minded enough to discuss the topic online. It's also important to note that the grandparents *"farmed* the *land,"* indicating the crash site property was likely for growing crops, not livestock, which were and still are commonly raised in the area.

If this remarkable "AllSouls" posting is to be believed, then it was this same hardworking 1940s farm couple - his grandmother and grandfather - that were the ones at home that evening, perhaps eating dinner or even getting ready for bed when the noisy crash-landing took place outside. Farmers are notorious for going to sleep early in order to get up for early morning chores but they sure didn't get to sleep early - if at all - that night. The agrarian duo were probably the ones who first stepped outdoors to investigate, likely in the afterglow of sunset and early darkness, and saw the metallic wreckage and bright flames, which were initially described as a "fireball" by Huffman when recalling the talk he heard around the crash site later that night. Deeply concerned, the couple then scurried back into the house and hurriedly called the authorities. Or to at least notified *someone* to go and do so, since not all rural farm families owned telephones in those days.

Obviously A.S. had also been talking over the years to his aging grandparents, who had let slip a few facts from that traumatic night, but evidently little else. They were probably the ones who intrigued him with the tale in the first place, long ago, or their offspring did. The grandparents "are deceased and were very hesitant to discuss where the crash site occurred," he admitted online. It's safe to say that the poster's grandparents must have had built up quite serious, solid reputations to have had their wild-sounding saga believed by their family and felt worthy of further detective work by the grandson. Evidently by 2013 A.S. was much freer to investigate the matter after his grandparents had passed away, and promptly did so. If the elderly grand-couple wouldn't impart details, he'd go to the region and find someone who would. It was that potentially historic and important, exciting and mysterious. It still is.

It is possible that A.S. learned enough about the farm's exact location to go there and speak to the current owner of the property in question, or perhaps to someone who had owned it for many years and yet stayed in the region, but not at the farm itself, perhaps now residing with relatives or in a retirement home in the region. It seems a little unclear however whether ol' "AllSouls" himself actually pinned down the precise crash location while poking about there. A.S. certainly didn't brag with rich details that he had discovered it, or claim that he had dug up his own souvenirs from the site, indicating it might remain a mystery, or at least off limits, even to him. Perhaps this also shows us that today the farm is a very sizeable and changing agricultural landscape and that even A.S. - and presumably all other visitors - are not welcomed to just waltz right in and start poking about for alien artifacts.

Sadly, there were no further *Topix* statements by the mysterious sources "Just Wondering" and "AllSouls." Nor were there any communiques to the postings asking for more information that were placed on the site in the summer of 2013. The matter became yet another sore point of frustration in fully exploring and explaining MO41. The two main forum contributors cared enough about the subject to participate in the story's message board threads, but not to fully answer questions posted and then not at all to requests for further on or off-line communication.

The overall allegations by the unidentifiable and apparently unreachable internet pair are admittedly exciting and sound very plausible, yet we must remind ourselves again that anyone can post wild stories and claims on any matter on the internet and not be held accountable for truth. Frauds and pranksters sometimes enjoy pulling stunts to get attention or purposely mislead people just for kicks. Government disinformation specialists allegedly muddy up waters in UFO investigations with false claims. Could any of that be the case here? Anything is possible, but the almost off-hand manner in which the information was displayed, and the detailed, authentic-sounding claims within the postings make it *appear*, at least, that these are two people who are very candid and genuinely helpful to seekers of truth. And thus their riveting assertions need to be thoroughly investigated. More updates are very much needed, and would be very much appreciated, to say the least.

Further sound detective work will be needed, online and on the ground. Due diligence is always required, and the difficult work of searches of farm properties between Chaffee and Scott City, just south of Cape Girardeau, will have to be undertaken based on these remarks. The only problem is that not all of the directions and locations and clues can be accurate. Using proper "Grammar" to sort out the wheat from the "Chaffee" will evidently be the hardest part in farm country.

CHAPTER FOUR

The Linda Wallace E-Book

"I picked up the bodies."

In early October of 2013, finally an e-book about the Missouri aerial crash came out for the public to digest at length. Pieced together by former Sikeston resident Linda L. Wallace, *"Covert Retrieval*: *Urban Legend or Hidden History"* offered some fresh new angles on the much-fascinating, multi-faceted MO41 incident. An online manuscript that is only offered via kindle through Amazon.com, Linda's web tome proved to be an unusual collection of enticing tales from old Wallace family history spliced and spiced with various possible governmental UFO-related cover-up factoids.

Working from her Ohio home, author Wallace spent several years toiling at *"Covert Retrieval*," but only stumbled into the UFO crash story by accident. Originally a Sikeston, Missouri, native, Linda was simply working on some genealogy software while trying to piece together her family's history, often to record older Wallace generational reflections before it is too late. The happenstance was that there were older family ties to both Parks Air College and the Missouri Institute of Aeronautics, which operated in 1941 out of the Sikeston airport in southern Scott County, the flight schools interacting closely. The upshot is that many uncovered old angles and riveting hints indicated that some of the Sikeston MIA flight training group - including Linda's own father, *perhaps* - were likely *the* "Army recovery team" that traveled one night to farmland near Cape Girardeau, to take over the ET crash site in mid-April '41. They evidently scooted south with the recoveries to Sikeston and likely held onto it all for a while at the airport's military base there. Then they apparently handled an often troubling aftermath for months and perhaps even years to come. In 2014, Linda offered interested parties a chance to view for free an early kindle edition of her new e-book so they could assess her MO41 investigation and judge for themselves its merits. Several small stories and compelling clues within the overall saga stood out and are worthy of further exploration and summation herein.

According to contemporary interviews L. L. Wallace conducted with the elderly matriarch of Linda's family - we'll call her "Maizey" - strange (and sometimes downright scary) events first took place in the Spring of 1941, just after her return to Sikeston following her graduation from another town's Christian college. In a recorded interview, Maizey remembered hearing rumors of a bizarre otherworldly crash event happening just weeks before her arrival in town, fitting our target date of on or around April 12th, since students traditionally graduate colleges in early May. The unusual incident was still the talk of the town for some, but often in hushed tones. For others it was a rather frightening mystery better left unresolved and undiscussed. Maizey was genuinely intrigued, as anyone with imagination and curiosity would be; her MIA fiancé (and some other airbase acquaintances) knew plenty and might have let slip at first with a few kernels of truth.

Evidently the titillating crash happened not far from a farm where as a healthy, vibrant young woman Maizey attended a friendly picnic that gorgeous springtime. The social gathering site was on the party farm's unique flat-topped hill, surrounded by lovely, level, lower countryside and farmland visible in all directions, *perhaps* in the general vicinity of the unique aerial accident. During the festive picnic, Maizey recalled, she pondered talk about the recent crash-landed craft "with little people" found dead, all while she stood on the top of the hill, looking about. She thought to herself half-seriously how pristine and serene the lush Missouri farmland appeared that this sight may have been an irresistible lure for outer space people to come see for themselves too.

The eye-catching picnic site was described in the Wallace book as about twenty-five miles northwest of Sikeston but southwest of Benton, Missouri, not too far from Chaffee and the nearby, smaller Morley. These might seem at first like some confirming clues to the actual crash-landing locale, but they just don't quite fit with what "Just Wondering" and "AllSouls" have stated, along with a few other 2013 *Topix* descriptions (see previous chapter), and still other sources. The picnic locale as a scenic place near the MO41 crash scene doesn't seem to add up in other ways, either; it's simply too many miles south of Cape Girardeau as remembered and described by Reverend William Huffman.

Decades after the picnic event, Linda L. Wallace traveled to the region and searched diligently for her family source's farm picnic site and then the 1941 UFO accident location, but came up empty, adding to the notion that the social gathering was not really staged within sight of MO41 events after all.

At any rate, our subject, Maizey, was a fresh-faced popular grad back then, engaged to a respected local MIA member we'll call "Randall." This happy couple was apparently writer Linda Wallace's mother and father although she did not specifically identify them to readers as such to protect family privacy, so that policy will be generally honored herein.

What *can* be said is that after her book came out, Linda described her father as "not a pilot" and technically "not in the Army" in mid-April 1941 but that he was a very able, educated young man possessing a solid knowledge of airplanes "and how to keep them flying." He was more of a trusted "flight technician" and "communications expert," a respected graduate of 1940 Parks Air College of Cahokia, Illinois, and someone familiar with the school's esteemed president, Oliver L. Parks. Mr. Wallace earned a "Bachelor of Science Degree in Aviation Operations" and became a trusted still-civilian member of the Sikeston airport's MIA flight training program. He was apparently one of its more reliable "line service" operators. More than just a mechanic, he was evidently a very trusted "crew chief" who might be called in today's world a "flight technology engineer and maintenance coordinator." Wallace was also part of the Sikeston MIA's "FTD," or "Flying Training Detachment," a uniformed unit that sported a special airplane "309 FTD" badge worn on their upper sleeves. Linda's dad sometimes wore a uniform, and sometimes a sharp suit. Unfortunately Mr. Wallace never seemed to speak to anyone in the family (beyond a few tiny tidbits for his fiancé-turned-wife) about the secretive MO41 affair, and moved his young family to Ohio a decade after the shrouded incident cooled off and faded from memory.

Before we dig into Linda's enticing e-book details, let's review some background. The Sikeston airport-based facility for the Parks Air College students, and pupils and associates of the Missouri Institute of Aeronautics, opened its hangar doors for its first semester in September 1940, operated by PAC/MIA founder and president Oliver Parks of St. Louis. Oliver was a friend of both Freemasons Hap Arnold and Charles Lindbergh, and half-named his fledgling airport near East St. Louis after himself. That was where Oliver ran his own PAC and coordinated others, with General Arnold's advice, in training young civilian pilots in all manner of flight. "Randall" got his initial schooling there, as mentioned.

Just a month after MO41, Oliver named the Sikeston airfield "Harvey Parks Airport," surrounded mostly by flat cotton fields. Harvey was Oliver's late brother, having died in a fairly recent, mysterious accident of some kind. The two Midwestern air facilities largely operated under Oliver's control. Featuring a fleet of attractive blue training planes with yellow wings, in most cases, plus dozens of eager student trainees and cadets, the PAC schools proved successful and growing at a time when America's potential involvement in the ongoing world war seemed to be inching closer every day. Some young men were pilots in training; others learned mechanics; and still others were educated in communications. A few other students worked in the PAC office, keeping the books and schedules, plus organizing and caring for the men and machines, maintenance and materials. Some worked in the 309th FTD, too.

In charge of keeping the whole Sikeston place well-stocked with tools and parts, office and p/x supplies, uniforms and meals was Ben B. Schade, the base "purchasing agent" and brother of Cape Girardeau County's new sheriff (see Chapter One).

In looking back, it must be admitted that at times the blurring of the lines between the Sikeston PAC and MIA organizations gets confusing. For instance, as mentioned a few PAC members sometimes wore civilian attire, and at other times Army uniforms. They sent graduates to the MIA who were sometimes considered full-fledged members and other times mere "associates." Civilian PAC grads who signed up for the MIA were removed from draft eligibility, yet were associated with the U.S. Army. In Sikeston the 309th Army air forces' MIA and FTD trained only mid-sized young white males - no racial minorities or women allowed - in the next steps of flight training and mechanical engineering beyond the rudimentary skills initially learned at Parks Air College. {The Army was still segregated in those days.} Flights commonly zipped up and around the Sikeston/Scott County skies and then back down during daylight hours only; there were no runway lights or nighttime air exercises at the facility, which sported a grass field for a runway, but with a paved area near the hangars. Accidents were actually pretty rare, a reflection on the expert education and care the PAC/MIA operation carefully laid out.

The Sikeston airport's MIA operation with PAC ties at the time of MO41 was under the command of Captain Charles B. Root, consulting often with Oliver Parks, with the civilian PAC in Sikeston run by A. B. Woodbury, again working under St. Louisan Parks. Captain Root used to work for Army bigwigs in Washington D.C. However, newlywed Root was transferred south to San Angelo, Texas, a few months after the alien crash, supposedly to command a new army-associated flight school there. C. B. Root went on to a very fine, well-rewarded career in the Air Force, achieving the rank of Major General. He even became at one point a special "Chief of Congressional Liaison" at Air Force headquarters in D.C., but it was his time as the Deputy Chief of Staff (from 1949-1951) at Wright Field's "Air Material Command" - long rumored to be an exploratory lab for captured alien technology - that captures one's imagination and attention the most, regarding MO41 after it occurred. Was it any coincidence Wright Field was a very likely site for the U.S. military's secret scrutinization of the MO41 materials and rumored other downed ET hardware? And that Randall apparently ended up there as well, years after MO41? Additionally,

Captain Root's eventual 1941 replacement at Sikeston's air training facility was Ralph C. Rockwood, who was hired in 1940 at MIA and given the title "Assistant Air Corps Supervisor." Ralph was still there at the time of the shocking April '41 UFO incident. Rockwood too "just happened" to end up at Air Material Command at Wright Field (around 1954). As a coda to their career connections, Root and Rockwood retired within days of each other in 1966.

In returning to Linda's "*Covert Retrieval*" more personal storyline, Maizey had heard part of the noteworthy Sikeston scuttlebutt going around that some men from the Missouri Institute of Aeronautics (and perhaps the 309th FTD) had been a direct part of the rumored weird crash's recovery operation, from weeks earlier. Tight-lipped Randall was not helpful in explaining or dismissing the unusual rumors; he told Maizey next to nothing. It apparently was a frustrating sore spot in their otherwise loving relationship, and wedding plans for the attractive couple slowly moved forward.

The local chatter continued on and off, in and around Sikeston for the next few years. A specific slice of this sketchy gossip in 1942 was worrisome to then-newlywed Maizey; speculation arose then that her handsome new flight expert husband had somehow lured, or directed, or even *caused* the alien crash – in northern Scott County, south of Cape Girardeau - due to his indulgence in an acquired military skill: using a heliograph. Author Wallace explained that a heliograph was a mirrored device used by American troops to signal each other in battle, utilizing reflected sunlight to send Morse Code-like flashes. This wild Sikeston causation accusation about Randall was of course utter nonsense, that Maizey's upstanding spouse had been somehow attracting or distracting alien beings in their passing spacecraft, causing them to fall to earth. Maizey asked Randall about this specific gossip and he dismissed it without great effort. until he spoke to his MIA superiors about the situation. They then hastily convened a meeting and apparently pressured Randall to tell his new wife to forget the entire matter and never mention it again. Randall passed the word to Maizey and she dutifully did as she was told, but that didn't stop others from wondering and gossiping. It was obviously an important issue that struck a nerve with the PAC/MIA and also the ranking 309th FTD training officers in charge, and possibly from those even higher up. Between Maizey's arrival from college in '41 and her marriage ceremony in '42, Randall was suddenly called away for substantial "advanced training" in Chicago, Illinois. For some reason, Maizey suspected the enigmatic area crash event was somehow responsible for this special, unexplained assignment. America had entered World War II during this period, and thus silence was important in all military-related matters. Randall dared not breathe a word of his new assignments. Maizey never did learn precisely what her man was doing in the Windy City, where covert operations were going on at the University of Chicago's special metallurgical laboratory on the top secret "Manhattan Project," the creation of the atomic bomb that was based at least in part on calculations for splitting the atom once worked on by the famous scientist, Albert Einstein.

Perhaps just a coincidence, it should be noted that a later-leaked Oval Office memorandum shows that President Roosevelt ordered a top American scientist "and Professor Einstein" in early '42 to search for ways to infuse ongoing atomic research *with the technology recovered from the nuclear power plant within "these new wonders," or "celestial" items that America had suddenly gotten its hands on. A.k.a., the crashed MO41 metal machinery and its possibly atomic power source.* But this does not mean that the famous physicist Einstein was actually told *where* the new atomic technology he was reviewing came from. A secret document by an Air Force general in 1947 mentioned that this shocking, lofty scientific goal *was* accomplished (see Chapter Three). It was all quite possibly involving some of the very recovered technology that Randall might well have seen and inspected, hours after it came crashing down. More likely for Randall, however, was that he was transferred to the Army Air Force Technical Training Command School, in Chicago. Still, Maizey's hunch may have been correct overall, yet her sweetheart's specific WWII actions in northern Illinois remain like almost all other angles of the cosmic Cape Girardeau crash: a fascinating mystery.

{A leaked document, now on Majesticdocuments.com, dated June 1947, weeks *before* the Roswell UFO crash, was allegedly co-written (or co-dictated) *by* Albert Einstein; it was all about the possibilities and ramifications of sustained alien contact with humanity in the future. "The presence of unidentified spacecraft flying in our atmosphere {sic} is now, however, accepted as de facto by our military," Einstein stated matter-of-factly. The remarkable document was annotated on the final page by Dr. Vannevar Bush, interestingly, but *still* doesn't prove Einstein was briefed on MO41, which was not mentioned in the great scientist's paper.}

Perhaps it is no coincidence that all of this intrigue in the Chicago area was under the surveillance of the Army's Counter Intelligence Corps, Linda Wallace uncovered. The CIC was operated by General George Marshall, who may well have taken charge of MO41 and another similar operation in early 1942 (again, see other chapters). The elite Counter Intelligence Corps conducted their own training school in the Chicago area. Wallace also discovered that the Army CIC provided security for the aforementioned Manhattan Project's nuclear research at the

University of Chicago. The various loose threads seemed all tied together, at least loosely in a sense.

When Randall finally came back to Maizey and the slower pace of life in Sikeston in late 1942, he seemed like a changed man. He often brushed off suggestions to go out and socialize, preferring instead to stay home and study – of all things - Einstein's theories on a nuclear physics and the nature of the universe. Randall had never done so before, but now he was more serious and introverted, Maizey recalled to Linda. There was of course a war on, and in '42 it was not going well for the United States and its allied world partners. To Maizey's chagrin, Randall was no longer as fun-loving and carefree in his off-hours. He continued to help train and guide flight trainees at the Sikeston airfield's 309th detachment, from late '42 into some of 1944, but Randall seemed more, well, inwardly sealed off. As if he had a huge secret locked inside.

To make matters worse for Maizey, a series of bizarre and untimely deaths had rocked the Sikeston community. She heard of some these shocking events through her friends in the MIA community. Many of the tragic passings were recorded in obituaries that author Wallace found in microfilm copies of *The Sikeston Herald.* Some of the local victims of sudden, suspicious-sounding "accidents" and "suicides" in Scott County were somewhat known to Maizey and/or Randall. One indigenous man, for example, was described as suddenly hanging himself in his barn after having "worked on a government project," the newspaper reported.

Every few months, it seemed, brought news of another alarming fatality. A hanging, a gunshot murder, repeated strange automobile mishaps. at times life in the Sikeston area was seemingly getting pretty scary and intimidating. Included in this troubling period was the shocking January 1942 lynching and burning of a black man by an angry, racist Sikeston mob, evidently led by two men from the MIA - yet no one was ever arrested or brought to justice, even by a grand jury. There was an effective stonewall, or "conspiracy of silence" involving both townies and some MIA men, to never speak of the horrifying murder. It proved one thing: Sikeston people could in general keep a secret in an atmosphere of fear (of retaliation), involving some local military personnel. Little wonder that some residents began to move away, perhaps including the owners of the crash site farm - the Scott County grandparents of "AllSouls" quite likely - and who could blame them? It was getting pretty unnerving, all of these incidences, and few were eager to get involved in more local mysterious events where so often someone ended up dead, by accident or natural causes or not.

One particularly disturbing newspaper obituary in the mid-1940s was that of a twenty-eight year-old woman who was married to an MIA associate that Randall and Maizey knew. When Maizey mentioned to friends at a card party this specific worrisome, unexpected death in a trend amongst others over the past three years, the young wife received a rather harsh warning. Don't talk about it "lest you be next," replied an acquaintance. This rather cold advice likely sent a chill down *everyone's* spine. Something *very* serious was going on, something that could not be dug into by locals or even the spouses of military men involved, and if someone asked too many questions. well it was best not to even think about it. And to keep one's mouth shut. Yet it was only human nature to privately stop in quiet moments and ponder the situation and its repercussions. Was our planet being visited by other life forms? And was our own government going out of their way to keep this fact covered up, even by committing outright brutal *murder*? Had some of these regional folks known too much, or talked too much, or kept actual evidence that was discovered and ruthlessly quashed by the military?

One specific name stands out today in this particular *Sikeston Herald* May 1944 obituary: "Fisk." Sound familiar? That was the dead woman's maiden name. A mere coincidence? Or a direct tie to the infamous Walter Wayne Fisk, the enigmatic Army intelligence officer (allegedly) who stole off with the Huffman family photo of a dead alien from the crash (see Chapter Two) and lived shrouded by secrecy until 2012? No evidence has surfaced as yet to directly tie the late Sikeston woman to the family of the mysterious Walter Fisk, but who knows? Stranger things have happened, and often within the MO41 saga strange things have *really* happened.

Linda L. Wallace found that the Sikeston newspaper's article on the MIA associate's spousal 1944 death referenced the selected pallbearers at her funeral. One of them was the name of a Parks Air College man - "Ray" - who some years later informed a few people that he had once been involved in the unique 1941 crash's quiet post-recovery transportation operation (see Chapter One). Ray had supposedly openly claimed he "picked up the bodies" - at the Sikeston airbase? - and presumably helped fly them out some place for the Army, perhaps the East St. Louis/Cahokia school where he was based, working directly for Oliver Parks. Now Ray was picking up, in a sense, another body, this one his friend's deceased wife, for her burial.

To sum this up: supposedly back in April of 1941 Ray was not a pilot, yet called in to the alien saga from his (still-mysterious) position at the Parks Air College in Cahokia, specifically to assist handling the loading and lugging of the recovered lifeless aliens. Therefore he flew aboard a sizeable plane down to Sikeston's airfield, where the alien bodies, and probably the spaceship itself, were carefully loaded at a secure Army-utilized hangar and piloted out. Was Ray in reality a well-trained, experienced medic selected to go in general to "pick up the bodies," or did he mean he was merely part of an impromptu security team, literally ready with "body bags" to bundle up and pack the dead aliens for their flight? Linda Wallace did not learn all the facts, nor does she care to speculate or elaborate beyond what is printed in her e-book.

It should be recalled that the small flight training aircrafts at the PAC/MIA center in either East St. Louis (Cahokia) or Sikeston would not have been nearly large enough to zip and ship the MO41 items across the country. If the Sikeston PAC/MIA wanted to simply fly *just* the three small extraterrestrials north, to Oliver Parks' St. Louis home turf, they'd utilize a fair-sized local airplane and have no need to import Ray or other Cahokia PAC people at all. No, logic tells us they needed a fairly big military plane for the *entire* MO41 load, recovered spaceship included, and for that someone in power had to scrounge up such a rarer, larger aircraft, probably a big military transport. Someone with real pull.

As described, PAC leader Oliver Parks was good friends with high-ranking Army officer Henry "Hap" Arnold, who could have provided just what the situation called for. Unfortunately, Hap was out of the country, on a mission for FDR in England in much of April 1941. Hap might have been contacted by a trans-Atlantic cable or call, however. General George Marshall, FDR's top military advisor and General Arnold's friend and boss, was another possibility for overall running this covert operation; whether *he* was familiar with Oliver Parks is unknown. Who precisely imported a big enough plane for the secret mission remains unknown, but certainly GCM had the pull and Parks had the secure Midwestern airfields with proper military security to pull it off quietly.

If Ray's amazing ET transportation allegation was true, it makes sense to speculate that he was part of a carefully Park-selected unit, flying down from the St. Louis area with a pilot or two, armed guards, other medical personnel, and cargo handlers. Some were likely informed of the true nature of the mission, some perhaps not and just obediently followed orders. Quite possibly Oliver Parks was aboard the large aircraft too, naturally curious and eager to see the crash materials stored at *his* school in Sikeston for himself. But it would seem that he could not jump *immediately* onto the first daylight flight down to southeast Missouri to see things for himself, or even pilot a small plane himself on the morning after, as he was very likely first busy in charge of rounding up his East St. Louis PAC/MIA team and the larger cargo plane, plus any specific materials required, perhaps while fielding phone calls from high-ranking Army officers wanting specific, top security action undertaken.

To speculate further reasonably, the three small ET corpses might have been wrapped in Army sheets or blankets due to their pungent smell and deterioration, or were placed inside available Army body bags originally intended for dead soldiers on the battlefield. This was done either at the crash site or at the Sikeston airfield. Possibly simple caskets might have been located and purchased on short notice in Sikeston, likely by helpful Ben Schade and Captain Root too. Getting the larger sections of the retrieved spaceship on a truck to drive to Sikeston… and then on board an airplane might have been difficult for any military aircraft to handle *unless* it was a cavernous cargo plane or troop transport, with a wide metal door that slid down to the tarmac for easy ramp-loading access.

Linda L. Wallace discovered that for some reason, sometime in 1942, Ray was transferred back to the Sikeston Parks Air College as an "associate" to the MIA's 309th FTD there. He was stationed there for the next two years, likely working with Randall, his old friend from Cahokia's PAC. Seemingly they both knew well what crashed to earth near Cape Girardeau and what was imported to the Sikeston Army airbase in '41. Gossip about the spaceship crash in the community had not died down entirely; Randall apparently kept his mouth shut, but Ray did not.

Yes, evidently Ray was occasionally indiscreet in his shocking revelations about the MO41 crash retrieval, at the time perhaps but certainly some years later, Linda learned. Evidently a few of Ray's pals reacted quite skeptically to his wild-sounding 1941 alien transfer claims. In fact, some who were out of the loop and knew nothing of the crash - Sikeston personnel "turned over" often - felt Ray was just plain crazy, Wallace uncovered through some interviews. Loopy or not, many years later Ray would be the very elderly man she would go visit in a secured nursing home (see Chapter One). Adult Linda was diligently seeking answers to otherworldly questions aged Ray stubbornly refused to acknowledge, obviously changing his attitude towards the remarkable subject over the years, unless he was truly addled by advanced age and memory loss. He did, however, perk up quite visibly for Linda at the very mention of her Army officer father. "Your dad was my crew chief!" Ray exclaimed with a broad smile, showing excellent recollection for activities from some fifty years earlier. "That was so long ago," he added. Then he sank back into silence when pressed for details on the 1941 incident, the interview essentially over. Sadly, Ray died not long thereafter, but in looking back, his brief revelation tells us how revered Linda's father was. Ray smiled with pleasant 309th FTD memories nearly a half-century later, indicating Mr. Wallace was a good man and a popular leader while in his military years.

It seems apparent that some folks in the Sikeston region held no inside knowledge of the '41 crash story, but others did, Linda discovered decades later. For instance, she revealed she once interviewed "a distinguished citizen" who wished to remain anonymous. That person discussed learning of the bizarre crash on farmland north of the city and that when this occurred, good-hearted residents from the agriculture-based community formed small groups to patrol the area, fearing trouble but also seeking any further physical evidence of the aerial accident - or possibly others related to it - on the fertile ground. Some citizens claimed to Linda to have witnessed the actual flash of light streak across the sky when the landing took place in the distance, however such aging tales might have been describing a 1938 meteorite that apparently dashed through the region, or were perhaps were from still other streaky disc and/or light sightings going on at times in the area's skies.

Digging deeper for *"Covert Retrieval"* from her Ohio home - the Wallace family having moved away from Sikeston in the 1950s - determined Linda discovered a man she dubbed "Sam," whose grandmother lived in the general north-of-Sikeston, south-of-Cape area back in 1941. Sam volunteered an attention-grabbing story of his grandparent being a ten year old girl that spring, just before America's entry in the global war already going on. We'll call the grandmother "Judy" and relay her story from the point of the day *after* the incident. It seems Judy's parents took her to a general store where the family heard aroused local talk about some sort of strange crash-landing of a large object in their farming community. Or was it *three* objects? The conversation was actually about a trio of three metallic-looking "boulders" that were now resting on a property not too far away. Judy's family went to see the unusual sight, and they were not alone, nor disappointed. Others had gathered to gawk, too, curious, quizzical, and excited. The tallest of the three odd items appeared to be about six foot in height, and they all had rather rounded, shiny-smooth surfaces, albeit damaged at times. Could this have been a crashed alien ship, ending up in a triad of main sections from the force of the sudden impact, blowing smaller bits of debris around the farm site? Or was it from one initially-intact craft that was skillfully cut up (via an acetylene torch?) at the site by someone, perhaps the military the night before? The grandmother evidently did not know. At any rate, Sam said Grandma Judy vividly remembered that the objects "reflected sunlight like metal" and seemed to give off a strong, foul odor. Perhaps this kept local folks at a respectful distance, especially when other visitors to the community arrived a bit later: men in uniform (but at least one of them in civvies). The team had pulled in riding on a long, eye-catching flatbed truck.

Judy recalled further that her own father - Sam's great grandfather - was a tall, strapping man who was present at the crash site, staring at the otherworldly objects along with many others. There was by afternoon still a considerable crowd, quite naturally piqued, and the rubbernecking grew worse when the apparent soldiers pulled in, mostly on their unusually elongated, flat vehicle. Author Wallace referred to this immediate area of local commotion as a rural "crossroads community" near the "Little River Drainage District." This hardworking agricultural base in the heart of Scott County presents a few problems for believability in the MO41 case, unfortunately.

Even L. L. Wallace admits the general drainage site of the grandmother's recollection falls beyond the "ten to fifteen miles outside of Cape" description given by Reverend Huffman. It's also so far south of Cape Girardeau that it seems pretty unlikely anyone in that sparsely populated rural community would call - if they even had a telephone - the Cape Girardeau police/fire departments in an emergency. They'd be much closer to Sikeston, or at least Morley, Oran, and Benton, the county seat. And worst of all, Pastor Huffman recalled the crash as consisting of bits of debris around one or two fairly substantial chunks of round metal spaceship, with dials, gauges, and weird writing inside, plus small seats, as if for children (by human standards). The wreckage was clearly an unusual aircraft of some sorts, with small gray alien pilot bodies nearby. "Gramma Judy" evidently didn't see anything of this sort, in fact she apparently saw or heard about *no* evidence of craft occupants at all. How she could have gone with her family to this site and not seen these seemingly unforgettable details? The discrepancies are puzzling, to say the least. It seems possible in hindsight that Judy's family shielded her from a sustained, close-up view, being an impressionable and possibly easily-rattled young girl. And also that a caring member of the community covered the stinky, decaying bodies with a sheet or blanket, out of respect for the dead - and the locals who came to gawk at the amazing scene.

At any rate, the "boulder" tale begs the question: *just how did the mostly-uniformed men know where to go that day?* The strangers rolled right in and took over the site, as if they knew exactly where to look, despite the fact there were precious few telephone lines (and thus phones) in the region and no established military outpost immediately nearby. Were they PAC and MIA personnel? The elderly woman could not say, looking back, but when there as a youngster with her family, Judy recalled, there were "about fifty people" on the scene that day, including the military men, roaming about, assessing the situation carefully, feeling the weird objects were likely going to be heavy and difficult to remove. Of course, these "boulders" were probably not large rocks at all, but perhaps to a small young country girl observing rather naively at a safe distance, the grayish-silver metallic objects appeared very much like huge stones.

Apparently no one had the courage to initially try to push or lift the odd objects after inspecting them, perhaps owing to fear and/or the considerable stench something gave off amongst the shiny items. A man "in business attire" appearing to be in charge enlisted Judy's father to help hoist the weird "boulders" onto the flat truck, and the muscular man obliged, paid for his efforts by the man in the suit. It turned out to be unnecessary, however, when the strange materials proved to be quite surprisingly lightweight and easy to lug. This detail, by the way, sounds rather in tune with Reverend Turner Holt's described experience at the Capitol Building, with Cordell Hull (see Chapter Five). Recall that Hull smilingly asked Holt to try to lift one of the sections of recovered spaceship and it proved surprisingly easy; the sub-basement room's UFO pieces were also glossy, smooth, and grayish-silver, just like what Sam's grandmother said she saw.

At any rate, Grandma Judy remembered vividly that her beefy father returned home after helping the military heft the objects and washed his hands repeatedly, unable to get the strong odor off. What would cause such a foul, lingering stench is unknown, but certainly the presence of rotting dead ET cadavers nearby - not mentioned by Judy - certainly comes to mind. Recall that it was unusually warm, windy, and sunny that Easter '41 weekend, factors no doubt accelerating any decomposition process.

Strange UFO activity had been witnessed by some local citizens in the skies prior to the crash event, Sam claimed Judy recalled in the pages of *"Covert Retrieval."* Weird "orbs" would supposedly fly around, break into smaller pieces, then later reform and dazzle onlookers. What precisely was going on was - *and still is*, according to some occasional regional UFO reports - is unknown, but it certainly does *not* sound like meteor activity, airplane maneuvers, or weather balloon behavior. Was it all a simple meteor or asteroid that hit the 1940s farmland and imprinted itself in not just the ground but in Judy's memory? Sam told Linda how he carefully showed Grandma Judy pictures of meteorites and she shook her head negatively, saying that each one was not at all what she witnessed that remarkable day in her youth. Meteorites the size of "boulders" would have done a considerable amount of damage on the ground, had they struck earth as such space rocks do, but no such destruction was reported by Judy.

{Others in the Missouri crowd watching the recovery process might have been the grandparents of a man we'll call "Jerry." This was another source of information within Linda's book, a man who said his own kin were present at the MO41 scene at some point. Jerry was born in Cape Girardeau in 1938, so their own farm was likely near town. Jerry was only about three when the incident occurred and thus too young to retain memories and understanding of what took place. But his relatives sure did. Jerry's grandparents were farm sharecroppers who moved to Minnesota in 1943 to work on a farm there and took the rest of their family with them, Jerry included. As he grew up, Jerry at times heard his grandparents talking to other relatives about the strange crashed object on the ground in a farming area somewhere south of Cape, north of Sikeston. The family would ask each other precisely what they thought it was that had landed, and where it came from. Perhaps from outer space? No one had definite answers, or at least would share them with Jerry.}

To their credit, Linda's discussion with Sam about his grandmother's memories came to a conclusion without wild conspiracy theories about alien visitations and ulterior government cover-up motives. However, it is very difficult to analyze Sam's unusual tale without coming to the conclusion that what took place in his grandparent's community was the busy "day after" the MO41 crash, featuring the Army's *second* recovery process.

To speculate, this might have been Sunday, April 13th, 1941, with Judy's family going to the regional store on that unusually warm Easter morning, their weekly day of rest, perhaps on their way to or from church. They then heard the startling talk and were later led to the crash site by fascinated locals, who were innocently or jokingly calling the items "boulders" due to their rounded shape from a saucer-shaped construction of a now-downed alien craft. The U.S. Army - specifically some Cahokia PAC people and Sikeston MIA cadets and their officers - along with a civilian administrator or two came knowledgeably rolling and strolling into the site to take over with the flat-bed transportation vehicle *as this crew had been there the previous night,* inspecting the scene carefully with flashlights and headlights, assessing what needed to be done about it, and picking up as much debris as they could. Possibly they removed the deceased alien bodies at that time, and yet their foul odors still remained at the scene.

Civilian PAC people often mingled with - socially and business-wise - with military MIA personnel. A man in a suit - perhaps a civilian - evidently had a position of authority in the farmland recovery team, it was vividly recalled. Was this Oliver Parks, the PAC president, flown in from St. Louis/Cahokia? Or Mr. Woodbury, his vice president and top civilian at the Sikeston PAC office? Or someone else high up from the Sikeston PAC/MIA or 309th FTD setup?

Possibly Captain Charles B. Root - the commanding Sikeston MIA officer - or his underling Ralph C. Rockwood took small groups of onlookers aside to warn them not to speak of the interstellar crash event, perhaps both that afternoon *and* the night before, such as when lecturing a rattled, Bible-clutching Reverend William

Huffman. It also seems reasonable to suggest that Ben Schade was also imported from the Sikeston PAC/MIA office that night and/or the day after, partly since he was a local, familiar with Scott County roads and landmarks. So many others at the fledgling Sikeston training center were from out of state and likely fairly clueless as to what highways and signals to look for. Ben Schade may have also exchanged phone calls - or perhaps even radio transmissions - with his brother, the Cape County sheriff and former Sikeston resident. This could well have occurred at any time: just before, during, or just after the military's first or second visit to the unsophisticated rural crash scene.

It seems plausible that perhaps one or two PAC and/or MIA members had stayed with the objects at the impact site overnight Saturday to Sunday. One man could have radioed in to the 309th headquarters the next morning some directions for the selected Sikeston recovery team to drive directly to that day, in precisely locating the obscure farm property in question.

It also seems reasonable to speculate further that it took some hours to get a select handful of trusted PAC/MIA men assembled that evening and then again the next morning. If it was truly taking place on a Saturday and Sunday, some personnel were probably away on weekend passes, and others were merely out in town looking for fun - on a warm, inviting Easter weekend? - totally unaware of what was going on with what we now call MO41. Meanwhile the flatbed truck had to be secured in the morning hours and any needed materials were readied and loaded for the trip, some twenty-five miles north. This process seemingly would have required Ben B. Schade the MIA "purchasing agent" to know about or even coordinate one or both of the "need-to-know" recovery operations.

Upon the unit's sunlit second arrival at the ET impact site, the otherworldly items were efficiently swept up by the small recovery team and whisked away, back down the road to the Sikeston airport MIA base, where they were more than likely unloaded inside a hangar, it can be hypothesized. The smaller bits of disc debris picked up the night before were there, perhaps in a simple Army footlocker. But the larger sections of the ET craft just imported needed more space. If PAC/MIA airplanes were inside or parked in front of this hangar, they had to be taxied over to other locales, to make way for the influx of men and trucks now moving in and out. And for any incoming cargo planes that would land and haul it all away soon. This activity, and some probable cursory inspection of the recoveries by authorized, well-educated 309th FTD personnel, would have required someone accustomed to understanding topnotch flying machines. *What Linda's pappy specialized in.* If he wasn't in on the actual retrieval, he'd likely have been there for its arrival and scrutiny at the base. Was this being handled by long distance telephone commands from down in Mississippi, where Army Chief of Staff George Marshall was last seen, having inspected Camp Shelby near Hattiesburg, but was now evidently "off the grid" Saturday night? Was a big cargo or transport plane soon on its way to Sikeston to allow the federal government, or at least the War Department's leadership, to take over the recovered goods? Clearly, less speculation and more solid facts and related information are needed, but author Wallace uncovered a fascinating side story from someone in her family that might help fill in a bit of a blank in this Sikeston airfield handover process.

According to a Wallace "unrelated family history narrative source" who guardedly spoke to Linda claimed a parent revealed that one day he or she saw a very large airplane parked on the Sikeston MIA/PAC runway, apparently around the time of the celestial crash and its covert capture.

The notable vehicle was much bigger than all the other planes that normally utilized the facilities and was surprisingly well guarded by uniformed soldiers, presumably armed. This was most curious, according to the Wallace source's recall, as was the airfield giving suspicious-sounding orders for *no one* to approach the enormous aircraft without permission. If this eye-catching runway incident wasn't all about MO41, then what else was it? The jumbo aircraft might well have been the special vehicle that "Ray" (and Oliver Parks?) flew on, coming down from Cahokia. Yes, this is once again mere speculation, but reasonably so, for the noteworthy sight of the big transport or cargo vehicle and its remarkable security detail around it would only have made sense if it was full of precious materials unlike anything seen in American history, on its way somewhere else. If it held, say, mere routine medical supplies or clothes and food for a needy foreign nation, or simply weary American troops inside the plane, it would not need such noteworthy protective measures. Something really big was up, it would seem.

The Wallace family source (who worked at the Sikeston MIA) stated that no one could recall ever having seen an airplane of this grand size and importance - with the uniformed guards - at the small MIA/PAC aerial facility. With his brother Harvey having passed on, it was Oliver Parks alone who ran the PAC/MIA show, very much in touch with his other aviation schools and visiting them at times in person, according to a Sikeston newspaper in September of 1941. But even Mr. Parks would have taken orders from someone as high up as George C. Marshall and the War Department. So if Oliver was around that day of the huge airplane's landing and waiting in Sikeston, he most likely knew precisely what was going on, and might even have been directing the procedure himself as a passenger, or even as the big plane's co-pilot. And doing all of this for a *very* high-ranking officer in the U.S. military.

A Missouri-Illinois air route was pretty routine in those days for PAC and MIA personnel; they'd fly from Sikeston to St. Louis/Cahokia, and vice versa, fairly often, perhaps using the north-south Mississippi River as a handy marker or aviation guide. It was only about 150 miles, done in an hour or so, although for a drive by automobile or truck it would take much longer, since there was no great four-lane interstate highway system in place as there is today. Hence, for military pilots and support personnel, using an unusually large cargo plane to lug alien hardware around, as opposed to trucking the tarp-covered materials (even at night), or utilizing a train, would have made a lot of sense.

Much of a MO41 trail left behind could have once been within the official papers (long since removed) of Air Force General Henry "Hap" Arnold, the Freemason and good friend of both G. C. Marshall and Oliver Parks. As author L. L. Wallace found, Hap Arnold had an underling named Jo Chamberlin who had been studying and accumulating UFO reports in the 1940s, then called "foo fighter" sightings. The Chamberlin notes were found by researchers within Hap Arnold's personal papers, decades later, strongly indicating that not only did Hap know about the private UFO study, he authorized it and followed its progress. While there was no mention of MO41 in what was left behind for historians to view, the Arnold files could well have been purged of the highly classified Missouri crash tale.

Let's add a little background to Linda Wallace's e-book coverage of General Henry Arnold. In the spring of 1941 Hap was Chief of the Army Air Corps, often based out of Wright Field in Ohio, the "Air Material Command" where so many UFO ties and tales have emerged over the past decades, although to be fair, General Arnold was often out of the country on military missions. Special metallurgical research facilities were apparently created at the airfield near Dayton, opening in 1945, much of its new construction situated underground and out of sight, for efficiently testing captured "foreign technology," according to tales from a 2013 exploratory book on Wright-Patterson AFB, and from a leaked 1954 U.S. Army manual. Wright Field secretary June Crain mentioned in a recorded 1990s interview seeing weird, metallic UFO wreckage once *and* General Hap Arnold *many* times at the air facility, and that in the 1940s Hap discussed alien spacecraft "with engineers" at the base offices. And as we have seen, Sikeston MIA officers Charles Root and Ralph Rockwood also ended up working at AMC at Wright Field. And it is probably not a coincidence that *more than half* of the Sikeston MIA's class of new cadets in '41 were from the state of Ohio, likely tied to previous Wright Field training or recruitment in Hap Arnold's days there.

"Operation Paperclip" scientists from Germany – perhaps like Dr. Otto Krause, mentioned earlier? - wound up at the secretive Wright installation, supposedly test-flying alien-Army hybrid spacecraft, confirmed by intel officer Thomas Cantwheel (see Chapters Two and Seven). "Project Blue Book," the famous 1950s and '60s official U.S. Air Force investigation - some say *whitewash* - of UFO reports. well, take a guess where that was based out of? The Ohio base was literally named in print as a safe locale for alien materials in the Army "Special Operations Manual" of 1954. And it produced some detailed, informed gray alien computer graphics - leaked in 1978 - based on actual research into recovered ET corpses. Then there's the famous story of former presidential candidate, Senator Barry Goldwater, asking a prominent Air Force general about W-P facilities. "I said: "General, I know we have a room at Wright-Patterson where you put all this secret stuff. Can I go in there?"" The general was most upset at the question and adamantly refused. Goldwater summed up succinctly: "I think at Wright-Patterson, if you could get into certain places, you'll find what the Air Force and the government knows about UFOs." The conservative leader served in the Army air command in World War II, became a pilot after training school, and heard rumors for decades about what was covertly going on at the Ohio airbase. Goldwater became a U.S. senator in 1953 while remaining in the Air Force reserve until 1967. He developed quite a fascination with UFO stories and spoke about them publicly after he retired from government in 1987, until his death in 1998. Obviously he knew the somber allegations about the recovered alien artifacts within the "Foreign Technology Division" and the "Air Material Command" at Wright-Patterson and felt they were serious enough to make some inquiries, and discussed it all openly before he passed away.

So where does Wright Field fit in with MO41? Linda Wallace's father "just happened" to get transferred to Wright in the 1950s. Besides General Arnold, General Nathan F. Twining, author of the "White Hot Report" mentioning the MO41 discovery while documenting the Roswell UFO crash recoveries, "just happened" to be situated at times there. And upon his installment as President Truman's first Air Force Secretary, almost assuredly Stuart Symington was in and out of Wright Field in the late '40s and early '50s, along with continuing his apparent friendship with fellow St. Louisan, Oliver L. Parks and possibly Charles B. Root as well. Mr. Parks' pal C. B. Root worked in the late 1940s as "Chief of Congressional Liaison Division at Air Force Headquarters," according to his military biography, and likely visited the U.S. Capitol Building at times, down in its lower floors. In the next chapter we'll see the significance of that unique, historic place.

Thus it seems very difficult to conceive that General Henry H. Arnold had no knowledge of or interest in recovered alien crafts and beings during the 1940s, particularly MO41. Quite the contrary, ol' Hap more than likely knew a great deal about the fascinating subject and possibly cashed in on it after he retired in '46.

Civilian Hap Arnold teamed with Douglass Aircraft just after World War II, Linda Wallace discovered, and created "Project RAND," mostly based out of headquarters in Santa Monica, California (not far from where Walter Fisk lived for a while). RAND supposedly clumsily stood for "Research ANd Development," while one contemporary investigative website report on the new organization described it as having "intimate involvement in highly classified UFO study for the U.S. government. From the beginning the men of RAND knew much about saucers." RAND opened its doors in 1946 and set off in at least one determined direction: *fulfilling an ambitious plan to produce a special design for a "world-circling spaceship."*

Imagine that, Hap Arnold was plotting with others in the upper echelons of the aviation world to go into space with a special craft that could defy our gravitational pull. RAND wanted to not just launch this high-tech, manned vehicle into space, but to "conduct activities" there and then "de-orbit" to return the intelligent craft safely to earth, unlike the unstable rocket launch technology of that era. Why, wherever did they get that idea? Wherever did the fledgling RAND team get access to this kind of super-advanced aerospace technology as a template?

Further, why was Hap Arnold involved in RAND flight technology and the planning of all things aerospace, upon the completion of the world war? Why didn't he, for instance, retire quietly at age 63, having suffered a whopping *four* heart attacks in the last few years? Or go into some other line of work? What hardware was he and others at RAND excitedly basing their designs and plans on - for zipping up into space and then returning safely for future missions - if they commenced this company and its lofty goals in 1946, *before* Roswell? Hap Arnold, the trusted friend (eventually) of President Roosevelt, who we know now had his own secret space program plans. A new vice president at RAND in the late 1940s was none other than General Leslie Groves, former head man of "The Manhattan Project," at times in touch with Dr. Vannevar Bush. And to boil it down and bring it full circle, Arnold's longtime friend Oliver Parks naturally went into the aerospace industry too, at least of sorts. In the 1960s, perhaps earlier, Parks was active in the aircraft manufacturing business - once Stu Symington's realm - when it suddenly veered into American *space research and development projects*, author Wallace discovered. Such big and important contracts weren't awarded and contacts weren't created haphazardly out of the blue in those days; Parks had to have plenty of inside aerospace flight knowledge and connections in the know to have achieved his position in NASA-related programs.

"I believe the facility had to contact the next in command," Linda informed me in 2015, looking back on Sikeston airport's mysterious April of '41. She added, "For the civilians" at the Sikeston PAC/MIA, "that would have been Oliver Parks' home office in St. Louis." Mr. Parks undoubtedly felt the crash matter was *his* turf, his people involved - his facility as a holding center? - and now *his* responsibility. He was pretty close to Hap Arnold, and after '41 Oliver went on to open and run (briefly) a flight training school at Cape Girardeau's airport. Parks – and *perhaps* Arnold - naturally spent some time there, and to this day, there rests two large C47 transport planes, evidently donated by the American military, at the Cape Girardeau airport. *One is described as once being the personal transport plane of none other than Hap Arnold!*

The "connect-the-dots" puzzle now seems to be fairly focused, perhaps finished.

More similar facts and claims could be disseminated here, it's just some of the "hidden history" and exciting information covered in the pages of unflagging Linda L. Wallace's *"Covert Retrieval,"* a recommended tech device read for anyone wanting to learn more about various reports of UFOs and mysterious government operations uncovered since the 1940s, all of it concocted initially simply to weave together some family history. Instead, her dedicated efforts help us to piece together one of the greatest stories never properly told in human - and perhaps alien - history. Yet one question remains, however: *where exactly did the MO41 recoveries go after the Army swept them out of southern Missouri?*

CHAPTER FIVE

A Dungeon's Three Glass Jars

"These are creatures from another world."

Our stunning story now twists in the most unlikely of ways, in the most unlikely of places: deep down into the musty spaces *below* the basement level of a world-renowned building found on the East Coast of the United States. That's where some remarkable confirmation of the MO41 evidence wound up being viewed and described later by - once again - a shaken, humble Midwestern Christian pastor to his amazed family only.

Before we get to that startling story, one must stop and fathom an unusual quote from author Linda L. Wallace's "family history narrative" source. When pressed as to what happened to the southeastern Missouri crash materials after they were whisked away in the spring of 1941, Linda's reluctant matriarchal informant briefly replied, simply but intriguingly. "The flying saucer or whatever and little people" from its Missouri crash were taken away and "hidden in a dungeon, or something (like that)." A *dungeon*? What an odd, distinct answer! Someplace resembling a basement or sub-basement room well below a structure's more occupied upper floors, in other words. How exactly Linda's relative found out about the supposed next locale for storing the MO41 goods remains a mystery, as the source refuses to say more, but it is time to support

and explain this very unusual and specific claim. *A
dungeon, or a place much like it.* Hmmmmm.

We now travel to Washington D.C., America's capitol, at
the height of the memorable F. D. Roosevelt era. There's
another paranormal story that arose from that period, one
that was also not delivered until the 1990s and seems to
reveal "the other side of the same coin," so to speak. It
involves the tale of two sisters, who have discussed their
memories of what their late father told them in a typed
letter - unwittingly just like Charlette Mann. The siblings'
papa was an Ohio pastor born and raised in Tennessee, near
his low-key but highly intelligent and successful cousin.

One of the supposed pair of long-deceased eyewitnesses for
our secondary, eye-popping MO41 allegement is famous,
while the other is very obscure. One grew up to a life in
politics and international diplomacy, the other in religion
and small town community aid. But neither of the two
diverse, related men lived to tell the revealing tale publicly,
nor anything else remotely unusual or otherworldly. Still,
what they are alleged to have inspected quite downstairs at
the majestic Capitol Building in the District of Columbia
was definitely *not* from this world at all. For the inside
scoop on this section of our paranormal tale, we first have
to track back to north-central Ohio. That's where a small
town music teacher Lucille Holt Andrew left this world in
2009. Before she departed, she dropped quite a fascinating
bomb on historians and researchers. Lucille and her
younger sister Allene Holt Gramly - hailing from obscure
Ashland and larger Mansfield, Ohio - had stories to
compare and share, claims that could make the average
citizen scoff at first listening. The two elderly ladies went
public in early 1999 with as many facts as they could
regarding what their long-deceased father - Church of
Christ Pastor Turner H. Holt - confessed to them
individually some fifty or more years earlier. It was all

carefully typed up in a neat letter, dated November 29th, 1999, and mailed to a UFO research group. It was then duly researched, written about, and publicized only on the Internet for years to come. A family source, a niece, confirmed that the family claim is valid. No one has disproved or proven a word of it since, but it is apparently quite relevant to more fully comprehending our Missouri crash saga.

It seems that at some point during the Great Depression, long-married and respected Reverend Turner Hamilton Holt (1894-1960) journeyed from Ohio to Washington D.C. to attend some religious conferences. He also found some time one day to meet with his beloved cousin and trusted friend, Secretary of State Cordell Hull (1871-1955), a very famous and adept politician-turned-statesman. The polished and professional Cordell worked - sometimes not all that closely - with *his* boss, President Franklin Delano Roosevelt, following Hull's illustrious career in the United States congress. White-haired, often-ill Mr. Hull had at one time labored for many years on legislation and governmental issues in the enormous Capitol Building, naturally becoming very familiar with its layout and quirks. Cordell served an impressive eleven terms in the House of Representatives (within the years 1907 to 1931), and almost two years in the same building as a U.S. senator as well. Secretary Hull also at times testified before congress at the famous D.C. structure when he was part of FDR's cabinet, speaking on important world affairs that affected America in a rather dangerous and desperate time in its long history, from economic calamity to nearing all-out global conflict. When summing up in his autobiography his years working with and for President Roosevelt, Hull remarked that he was allowed to discuss every sensitive topic most candidly and received "frank answers to most secret matters." Books published years later on Hull and FDR exposed many such secrets withheld from public view for over a decade after their deaths. Theirs was a world of touchy and explosive issues, some kept clamped down, others gradually let out of the box, as it were. But one thing is certain: *Hull knew the Capitol Building quite well.*

Lucille and Allene explained that Cordell Hull was instrumental in getting Turner Holt an advisory position linked to the U.S. government, giving the dedicated Ohio preacher a sound reason to occasionally travel to our nation's capitol and visit the famous secretary. According to the Holt sisters, while on such a trip and chatting one afternoon at State Department headquarters, Cordell - perhaps rather impulsively - decided to ask his cousin Turner to travel across town with him to the imposing Capitol structure. The sisters were not certain of what specific year their late father related his version of this tale, but agreed it was sometime *before* America's involvement in World War II, and certainly before the terms "UFOs" and "aliens" were part of the national lexicon and mainstream culture, from 1947 onwards.

Cordell led Turner up to the esteemed building's steps and stopped. He cautiously made the evangelist take an oath, swearing to keep what he was about to see a secret, just between the two men. Turner agreed and repeated the oath, then Cordell led him into the building, heading down a series of stairs, far past the official basement level of the building. They walked further down into the seven-storied Capitol "sub-basement," where historians say there is actually *more* square footage than the famous upstairs floors. At any rate, the two men then viewed what certainly appears to be the results of the Army's retrieval of the Missouri crash remains.

There can be no debate or doubt that lower floor storerooms *were* utilized daily well below the Capitol Building's ground floor in the 1940s (and before then). They likely had the look and feel of a kind of *"dungeon,* or something." In 2002, a researcher interviewed the Capitol Building's curator, a knowledgeable, educated woman who confirmed that in 1941 its subterranean floors were indeed *full* of smallish sites and spacious suites, but described them mostly as *storage rooms*. Today they've evidently been developed and modernized, but in stodgy Cordell Hull's day, the below-floors sites were mostly used for carefully storing dusty items the government wanted secure and out of sight but not dismissed or forgotten.

In the 1930s there were armed, uniformed security guards - the Capitol police - stationed inside and outside the 541-room Capitol Building. Records show that on July 1, 1940, they were augmented by twenty very experienced new officers, on orders of the Roosevelt administration, thanks to general nervousness about possible Axis Powers terror threats as much of the world waged war. The new guards were handpicked from the FBI, the Secret Service, and the D.C. Metro force. So by the following Spring of '41, FDR knew full well his revered Capitol Building - with a huge American flag flying out front - was perhaps the best-protected building in town, if not the entire nation. Obviously this augmented Capitol Police force did not want to give a highly esteemed and internationally famous man like Secretary Hull any problems. They likely recognized distinguished, angular Mr. Hull quickly and let him in with little questioning, with Pastor Holt as his accepted guest.

A perplexed Turner Holt and a calm Cordell Hull did plenty of walking when they arrived at the Capitol Building. They found the proper lower floor and strolled down its long corridor, the visiting clergyman starting to tucker out. One of Pastor Holt's daughters recalled their father describing with a chuckle his asking Secretary Hull with exasperation, "*Where* are we going?" after the duo's walk became longer and more puzzling. Cordell evidently did not answer, but simply stopped at a specific door -

perhaps locked - and managed to get it open. The great statesman then reached in and flicked on the lights within the possibly mildly dank room. Cordell then stood back and let Turner enter and gaze around, somewhat like Alice stepping through the looking glass.

On the floor of this particular Capitol Building downstairs storeroom sat crates of metallic objects - silvery shards and shreds - and some large glass "jugs or jars," one Holt daughter tried to recall. There were a few larger and somewhat rounded, covered objects in the back of the room. This only brought the pastor a little more mystery as they gazed around: what in blazes were these strange things? *Pastor Holt somberly told his daughters that he was informed by Hull this was very real physical evidence for extraterrestrial visitation.* There were pieces of metallic crash debris from a damaged, circular, aerial "vehicle" (now in pieces), and the bodies of its nonhuman crew, and it was all quite top secret, Hull somberly assured Holt. He'd have to keep his mouth shut about it all, but here was a fabulous feast for the eyes - and the mind!

The unbriefed, unprepared Revered Holt had to have been utterly flabbergasted. The Ohio evangelist was allowed - even encouraged by Cordell Hull - to pick up and look over the shards of metal debris. He was then gently prodded by the Secretary to try to yank up the smooth, rounded edge of the downed ET disc, which he found to be broken in sections and covered with some sort of tarp or "wrap." Patient Mr. Hull gave one large chunk's leading edge a lift, so Pastor Holt did likewise. It was surprisingly light, he recalled later to his daughters, and with little effort the silvery ship segment flipped upwards in the rattled preacher's hands. He set it back down and stepped back, looking around, trying to soak it all in, absolutely floored. Evidently no one else was present in the room, and no one else came by to join the secret encounter. If such a third party was present, evidently T. H. Holt did not mention him. Overall it was the most staggering, amazing secret the simple clergyman had ever seen or learned, undoubtedly also for his cousin, the now-calm cabinet member. But the weirdest trauma was yet to come, within the glass objects present.

The shaken Turner H. Holt stepped up closer to inspect the glass "jars." Strange human-like little bodies were stuffed inside these clear containers, floating in a clear fluid, which Holt allegedly described as "formaldehyde." How could the evangelist have known this, since this special liquid is identical in appearance to water? Perhaps Hull informed Holt of this detail, or the pastor simply deduced it was this methanol preservative, used often in keeping biological specimens well maintained, long after physical death.

The unusual creatures in the fluid looked like gray, rumpled children with abnormally large, round skulls, at least at first glance. *"Three glass jars,"* the elder Holt daughter emphatically recalled on tape very firmly, shortly before her death, as to how her father precisely described the containers. *Three* was the specific, memorable figure clergyman Holt used, she said. The two elderly sisters gave an interview with Canadian researcher Grant Cameron in April of 2009, and the results can be heard online to this day, repeated for posterity by Lucille Holt Andrew (who died just five months later). *Three alien beings from one downed, round, damaged craft.* While speaking side-by-side on a couch in the 2009 recorded memories, Allene Holt Gramly did mention that she initially recalled her papa somberly informing her that he had once witnessed four alien creatures in glass jars. Then, Allene stopped herself and changed her mind. *"Three* or four," she recalled, thinking about it more clearly. This backs up Lucille's emphatic recall of *"three glass jars"* in the recorded interview. The sisters had indeed originally referred in their November 1999 letter on their recollections as *four*, but eventually realized they had been mistaken, overreaching by one jar/alien. Unfortunately, to this day some UFO web sites and stories still stubbornly and erroneously refer to the allegation as "four aliens in four jars." This is clearly not accurate. It was *three*. Three jars, three dead ETs - *just like near Cape Girardeau.*

The trio of unusual-looking dead alien bodies were afloat with "their eyes large but closed." This seemingly small detail strongly indicates the witnessed ETs had eyelids, another human-like feature that shows they were once very real creatures and not somehow doll-like robots or fabricated androids. Minister Huffman did not mention seeing the MO41 aliens with eyes shut, nor did Charlette Mann recalling her photo subject, leading us to wonder: *who shut them, and why?* Did they close up naturally on their own, after death? At any rate, each odd jarred and jarring entity was evidently identical to the next, and strangely undamaged. They were all estimated to be about four feet at most in length, albeit scrunched up a bit inside their glass containers, the humble evangelist told his daughters. They were gray-skinned, slender in shape, with long thin arms and legs. Big bulbous heads and evidently black eyes. Sound familiar?

"These are creatures from another world," Secretary Hull calmly and knowledgeably informed his cousin, or words very close that effect. This was clearly no elaborate hoax, judging from the serious Cordell's face and the inflection in his voice. Normally somber Hull was about the last man on earth to play practical jokes or waste time on someone's trumped-up hoax. *Aliens had landed, died somehow, and we had 'em. Their advanced airship too.* It was a huge, juicy secret, too hot to keep inside for the excited statesman.

The minister was undoubtedly numb, perhaps barely able to think straight. He took a closer look and noticed the three odd entities all had three long fingers per hand and one long thumb. They had no discernible ears, nose, or hair, as his daughters recalled him saying. It was likely a strong, creepy memory any person witnessing would never forget, for the rest of his or her life.

What Reverend Holt kept coming back to that day, evidently, was the size of the craniums. Those big, round heads, they were amazing! Almost hypnotic in their bizarre appearance; the two men allegedly spent some considerable amount of time discussing the implications of this particular disturbing fact in that special store room. "They talked about that for quite a while," one of the sisters recalled their father saying. Perhaps in hindsight Holt and Hull wondered and pondered if a bigger alien skull meant a frighteningly bigger, smarter brain, and thus a far greater overall intelligence than mankind's, apparently able to traverse the vastness of space with stunning ease in advanced airships. {Yet how smart were they if they had ineptly crash-landed?} Was this a bad sign of something to fear? More of these strange "foreigners" coming, looking somewhat like us, but weirdly misshapen?

If this wasn't the mindboggling MO41 evidence, kept in cold storage just down the street from handicapped FDR's White House, then what else could it possibly have been? Holt's description of what he witnessed is an impressively identical fit for 1941 Missouri's odd yield in every way. The exact number, size, shape, weight, and colors. all the independent Ohio details from that obscure D.C. Capitol underground room fit perfectly with the southeastern Missouri discovery/recovery operation. Clergyman Holt's story surely helps explain where exactly the MO41 artifacts were flown to after leaving the Show-Me State. Where else would a Washington-bound commander-in-chief who loved

Freemasonry order the dramatic, historic items taken? At least at first? Freemasons were wild about the Capitol Building. Furthermore, if D.C.-based General George Marshall was involved in commanding the MO41 recovery procedure as suspected, the Capitol Building - where he would soon have to testify before the "Truman Committee on Military Affairs" - is exactly the place where he could at times keep his eye on the nearby prize over the course of the spring of 1941. Marshall and Truman were also Freemasons, and many of them met in Washington, and today some speculate that a special masonic room or lodge existed under the top floors of the Capitol itself. This would be handy considering so many members of congress were, in those days, members of this secret society, including the sitting Speaker of the House, another FDR ally. "Pathway to the stars," that's a Freemason slogan still in use today, such was their obsession with the heavens, as painted into the soaring Capitol dome's artwork.

"We can't tell the American people about this as it would probably start a panic," Cordell supposedly informed a rattled Turner Holt in the dungeon-like Capitol Hill room. Hull likely want to both impress his cousin with the evidence *and* desired to gauge his reaction from a religious or spiritual man's point of view. He unfortunately didn't tell preacher Holt where precisely the materials came from or how the U.S. government got their hands on it all, but we can assuredly take an educated guess nowadays.

"Three glass jars." A key, informative phrase. If this was truly an accurate description of the containers utilized for the recovered Depression-era alien beings, it is highly indicative of the involvement of a hospital or a medical science laboratory in storing biological artifacts. Glass jars large enough to store entire bodies were probably pretty rare, but such items *were* created in those days, in glass factories that shipped them to academic and medical facilities. The big jars were used mostly in scientific research into the medical conditions of the preserved deceased, be it entire bodies (of animals or human children, most likely), or their individual large organs. This means someone very trusted within the army recovery unit made a quick trip to a lab or a hospital, searching for something to both contain and *show off* the evidence. The most likely site to quietly visit for this unique purpose was either Bethesda Naval Hospital or Walter Reed Army Hospital, both located in the D.C. area, where autopsies are regularly conducted, complete with a cold storage area for the corpses. Supplies, such as glass containers for bodies and/or organic body parts would have been there, most conceivably.

{One other source *might* have been the army base Fort Meyer, near Arlington Cemetery, just across the Potomac River from the White House and the Capitol, where many deceased soldiers were held for autopsy, then readied for burial on the grounds. It just so happened that the Fort Meyer/Arlington complex was home to General George C. Marshall, FDR's Army Chief of Staff. His official residence there was "Number One General's Row." The longtime serviceman dutifully rode his trusty horse on the grounds almost every day when in D.C., even in his older years. General Marshall was quite familiar with Arlington's cemetery and burial facilities due to his many decades of working with and honoring his fellow soldiers and veterans, attending services and funerals at the site. According to leaked documents, George may well have been *the* leading figure in taking charge of and storing the MO41 discoveries.}

One small piece of possible evidence tying the general to the great capitol architectural icon is a 1941 photograph, possibly from Friday, April 11th, '41, likely the day *before* the crash near Cape Girardeau (or possibly from a few months later that same year). At any rate, the snapshot shows General Marshall in his uniform, standing rigidly next to an airplane on a runway. The online caption reads *"George Marshall arrives at Esler Field"* near Camp Beauregard, Louisiana, to review that military installation and its soldiers' readiness for possible war. This might well have been his scheduled stop on his mid-April camp tour. Before this springtime journey GCM stated in a letter he would only be taking along "two pilots and Colonel Ward," his nearly-fifty confidential aide.

In the camp tour airfield photo, two uniformed Army officers are shown greeting the stiff-looking Marshall. But it's what is *on* the plane, not within it, that is the great curiosity of this exposure. Clearly adorned - perhaps painted - on the side of what appears to be the general's personal aircraft is none other than the Capitol Building's dome! Imagine that, Marshall was so linked with - or obsessed with? - that particular building that he had an image of it painted on his personal aircraft. Why? Since we know it is very possible the MO41 materials ended up *inside* the Capitol Hill complex, where the general was scheduled to testify in the weeks to come that April, is it really so hard to believe that Marshall - in conjunction with his superior, FDR - stashed the goods in the building's obscure sub-basement store room? Even if the remarkable GCM photo was somehow from another visit later in '41 by workaholic Marshall to the ongoing famous "Louisiana Maneuvers" war games in the area, the question remains *why the Capitol Building logo was painted on the side of his plane in the first place*, with no accompanying inscription or explanation. After all, Marshall worked out of Army offices in D.C., and the War Department building, and reported to the president at the White House. He did not really take orders from anyone in the Capitol's two

houses of congress. Yet the famous Capitol dome was his -
or *someone* high up's - choice of personal identification.
Also appearing on the plane's side nearby was a logotype
of his very rare four-star rank as a U.S. Army general,
showing all concerned who was in charge of this plane and
its base-touring objectives.

Interestingly, Reverend Holt did not mention seeing any
surgical scars or stitches visible on the three dead bodies he
viewed, so it is very possible that the trio of aliens had not
been given autopsies by that point.   Preserving their
remains in formaldehyde would presumably have been a
good way of keeping them fresh enough for just such an
investigatory surgical procedure later on.

Three *glass* units would have been utilized for best
displaying the alien beings within, and this was perfect for
any member of congress, the State Department, and/or the
president and his most trusted military aides to look over
the extraterrestrial biological entities.   Clear glass, clear
fluid inside, to best see the shocking evidence within.   It
strongly indicates that the creatures were already being
handled by trained *medical* personnel.   That they were *not*
being prepared for a burial, nor cremation.   A wooden box
or casket, or a darkened or smoked glass container *could*
have been utilized to hold the ETs, but obviously this is not
what the Holt sisters recall their father telling them.   They
were considered fragile specimens for further scientific
examination, we can surmise.

Another important factor: the three Capitol aliens were *not*
dry bodies stretched out, flat on their backs, for mournful
viewing as if they had been handled at a funeral home.
They were not "stuffed" or "mounted" in any way, as if
they had been long since taken to a taxidermist, as captured
prize animals are when treated when hunted for sport and
displayed later as dried-out or mummified trophies.   No,
these were bodies seemingly ready to be cut open and

examined, perhaps again and again, chiefly by different trusted and trained surgeons, zoologists, doctors of various medicines, and pathology experts. All sworn to secrecy beforehand, of course.

A great factor in supporting the notion that the MO41 materials were quietly trucked to and stored inside the structure in mid-April of '41 - perhaps within days of their retrieval by Army personnel - is that the United States congress was on a Passover/Easter holiday. The Capitol Building was particularly empty at this time of the year, with perhaps scant few visitors, members of the press, and government employees around at all. It was clearly *the* perfect time to covertly import shrouded alien cadavers and spacecraft, and stow them quietly in a sub-basement room. Since they were kept under wraps, who in the building knew? Perhaps it was also an ideal time for any military leadership, FDR crony, or trusted scientist to also slip in unnoticed and inspect the stored recoveries without detection or inhibition. Congress would not be back in regular session until Monday morning, April 21st. The place was nearly a ghost town for a while. The coast was relatively clear for importing a team of workers or soldiers to install the stunning prizes. It is mere conjecture, but perhaps logically this was also the same week Hull and Holt took in the Capitol's sub-basement secret.

In going public in the late 1990s with the amazing Hull-Holt allegation, Lucille Holt Andrew knew it would trigger some controversy, but in her case, the mainstream media pretty much ignored her. Lucille had no proof, not even a deposition or confession from either Cordell or Turner, or a tape recording or home movie of either man's account of the otherworldly evidence. It was just Lucille and Allene's word, basically. Before Grant Cameron's interview, in 2008 researcher Linda L. Wallace ventured to the sisters' Ohio hometown to interview the elderly women and a few other family members, but in the end could not find any sort of apparent fraud or deception on their part. Their riveting claim seemed reasonably plausible and unchanging, minus the original "four jars" to "three." It was a bold and courageous story to relate, one that would likely bring the Holt clan ridicule from some, in their community and from around the country. But both women said that Turner informed his two girls to go ahead and publicly relate the ET tale only after he was dead and gone, that it was simply too important for history to just forget. The two daughters dutifully did as they were told. They did not alter or inflate their claim.

History shows that tuberculosis-tainted Cordell Hull resigned from FDR's cabinet in 1944. He needed a rest, having held the position of Secretary of State longer than any man in U.S. history. He was tired, in troubled health, yet managed to hold on until 1955, dying at age 83. Before he quit the State Department, Hull helped FDR lay out great plans for the new "United Nations" that later came into being, with some observers later referring to Cordell as "The Father of the U.N." Meanwhile Turner H. Holt lived a fairly rich, full life, but passed away from leukemia in a Columbus, Ohio, hospital, at age 65. Both gentlemen were well-educated scholars, originally from impoverished parts of Tennessee, and had worked hard to achieve the American dream, often in tough economic times. Mr. Hull's accomplishments were so great, it's been emphasized that increasingly ill FDR offered him the vice presidency in 1944, perhaps to someday set Cordell up to become the next president and leader of the free world. He obviously would have been calmly knowledgeable about extraterrestrial visitation and other "foreign affairs" that other candidates for the job would not have been. Cordell Hull turned it down in order to slow his life down and rest in retirement, but he did win a Nobel Peace Prize for his life's work, toiled on U.N. issues, and went on to write his memoirs.

Meanwhile, surprisingly, Turner Holt went on to pen a book as well, entitled *"Life's Convictions,"* released a year after Cordell's death and now long out of print. Holt died 1960, ironically just months after Reverend Huffman in Missouri. Pastor Holt in Ohio left behind a grieving widow, Vina May Holt, the woman he loved and married many decades earlier; their two daughters waited until she passed away in 1993 before deciding to take any action on their extraordinary extraterrestrial story. Devoted and devout Vina was never interested in such matters and apparently didn't even approve of such talk. Once she was gone and a respectable amount of time passed, the two sisters were free to finally step forward.

{Note: a researcher's check of Hull's office "desk diary" did not show any visits from Turner Holt during working hours from 1938 to 1942, however it is obvious that as a visiting relative the duos' meeting would have been a social event in off-hours, not recorded.}

Lucille Holt Andrew was not entirely certain, looking back many decades, on precisely *when* her daddy told her the astonishing tale, or when the alien encounter took place. One Holt-claim researcher feels the stunning father-daughter conversation was held within a year or so (perhaps sooner) of the spooky Hull-Holt adventure, which makes sense, while the story was still fresh on Turner's mind. Lucille was either late in her teenage years, 19 turning 20 years old, *in 1941*. As mentioned, it has been said by some observers that the daddy-daughter chat had to have been held *prior* to World War II, which did not break out - for Americans, at least - until the final weeks of '41. The Capitol would have regularly been under some notably heavy guard *after* December 7th, '41, very memorable for the preacher, and likely not so readily open to a casual

Holt-Hull visit. Indeed, Cordell would have been far too busy for a Capitol Hill rendezvous like this, after war began. Thus we have a plausible time link to MO41, seeing as how that event took place in the early spring of the same year. It is admittedly conjecture, but. the macabre memories of the celestial machinery and debris and the weird corpses on display may have been so striking and traumatic on the mind of Reverend Holt that he urgently told his oldest, most mature daughter *within days* of his witnessing the staggering evidence, quite eager to get it off his chest as soon as possible, much like Reverend Huffman.

Allene Holt Gramly stated she was not informed by her father of the sub-basement Capitol encounter until 1948. However, she was the younger sister and evidently needed to mature first, and it has been hinted at by Lucille that even *she* really wasn't paying *that* close attention to her father when he unburdened himself of the tale, possibly seven years earlier than her sister. Teenaged Lucille's still-developing mind was on other things, she admitted decades later, and the notion of dead little aliens supposedly stored in a secret Capitol subterranean room probably sounded originally like her father was joking, or drunk, or perhaps merely testing her with a wild tall tale. Allene said she *tried* to listen carefully to her father describe the vivid, traumatic memory in '48, but a small child jumped up into her lap to get attention at that time and that ruined most of the rest of the ET conversation, which was sadly never brought up again.

Both Holt sisters look back with pride on their father's integrity and his infrequent advisory work for the American government, set up by his Cordell Hull connection. They proudly displayed an authentic 1935 letter that President Roosevelt sent T.H. Holt, on special, water-marked, White House stationary, thanking the minister for his participation in apprising the U.S. government on a small matter.

Although this really proves nothing, there was once the possibility in the family's mind that '35 was perhaps the year Preacher Holt went to Washington and saw what he later claimed, yet neither sister was really able to reliably pin down precisely when the Capitol Building excursion took place. Apparently there were a number of Washington visits for Reverend Turner Holt over the next decade, making 1941 more possible. The sisters' interviewer, Grant Cameron, set the possible year online as being 1939, acting on the vague suggestion by Lucille at one point that perhaps the event occurred "in the late 1930s." The yellowing 1935 document the sisterly duo held may well have had absolutely nothing to do with the actual Holt-Hull Capitol Building visit.

When told for the first time the Reverend William Huffman story of three aliens crashing down and being captured in southeast Missouri in April of 1941, Lucille Holt Andrew told researcher Linda L. Wallace that perhaps the Holt-Hull encounter down inside the Capitol Building "could have occurred later than she originally thought." Not the 1930s after all, but maybe in 1941, Holt's older daughter agreed. *And that quite possibly the two unusual alien sagas were indeed related*, the very same items, Lucille suggested.

To their credit, both Holt women said their father simply described the occupants of the crashed disc as "little creatures," not "ETs" as we would term them today. And that he referred to the spaceship as a "vehicle," not a "UFO" or "spaceship," as popular vernacular mentions them in the present-day. This certainly recalls the precise terminology that clergyman Huffman used when conversing excitedly with his family in Cape Girardeau. The saucer-shaped craft was "in pieces" and somewhat "silver" in color, Turner told his daughters separately, years apart. They re-emphasized that "he thought the ship would be heavy, but it was light." Holt oddly became a virtual parrot for Huffman, inadvertently; the stunned Ohio preacher ended up backing up and confirming the stunned Missouri preacher very nicely. At least, according to the two daughters of the Ohio eyewitness, speaking out in the late '90s, compared to the two granddaughters (Charlette and her unnamed older sister) of the Missouri eyewitness, speaking out in the early '90s.

The two Holt daughters were both informed by their father *not* to speak publicly about the claim with anyone while worried Turner was alive, and it soon became apparent that the simple pastor's own wife, Vina May, was not even aware of details at all, as she had no interest in such matters. Disinterested Vina's chief concern in life was to keep a respectable home and raise an honorable family in a small Midwestern town. In conservative times, blabbing about seeing aliens and spaceships in D.C. secrecy - while breaking a special oath of silence - was not at all considered acceptable social behavior, particularly for a respected Christian minister's kin in a sleepy community. Especially when Turner Holt undertook occasional advisory work for the government and wanted to keep his jobs, his income, and the prestige of hobnobbing with the famous Secretary Hull. It makes more sense, then, that Pastor Holt would take the time to privately inform his two children, since his spouse did not really care about such nonreligious matters, but then ask them to keep it to themselves until he passed away. To eventually pass the story down to the next generation, before his memory faded, that was critical to the eager-to-talk Turner.

The offspring of the two Holt sisters stated in interviews that their related parents were solid citizens and not prone to make up fanciful lies for attention, and that they believed their grandfather Turner's wild-sounding allegement. However not one scrap of physical, tangible evidence remains to support this remarkable tale, and in his memoirs,

Cordell Hull never let slip a single word that even hinted that the tale could be true. He of course could not for security reasons, no one could. The topic was undoubtedly highly classified and considered "Above Top Secret." Perhaps Mr. Hull too was sworn to secrecy, by the president or the recovery operation's military leaders, but if so he obviously broke that oath by showing the items to Turner Holt, who then promptly broke *his* sworn oath by blabbing to his offspring. Much like Reverend Huffman breaking *his* "oath of secrecy" to his family just minutes after getting back from the MO41 crash site the evening it happened. People so often naturally feel the need to talk, to share or purge, especially about something very unusual, exciting, and historic. It is only human nature to feel bursting with emotions and memories that one needs badly to get off one's chest, so to speak.

Over the centuries, several dead U.S. presidents and congressmen have been taken to the Capitol Building to lay in state before their burials at funeral services held elsewhere. In some rich irony, on the evening of Saturday, April 12th, 1941 (the probable date of the MO41 crash), Cape Girardeau's main newspaper, *The Southeast Missourian,* displayed a prominent front page b/w photograph of - of all things - a dead body being carted out of the U.S. Capitol Building! A United States senator, Morris Sheppard of Texas, had died just days before, and a horse-drawn carriage was ready outside the building to ceremoniously lug his casket away. This was a newsworthy image splashed all over the nation, including the Cape Girardeau region. One or more representatives of the Roosevelt administration were undoubtedly present in that funereal process, paying their respects inside and out of the Capitol, viewing the somber process played out in public. Then it was off to a holiday congressional recess. So as we can see, *storing corpses inside the Capitol Building was common practice, and would have been fairly easy that Easter/Passover week, with almost no one around.*

Is it realistic to think an entire spaceship could have been hauled into the Capitol structure without notice? When one ponders the situation, the answer grudgingly becomes "*yes.*" With the holiday evacuation of Capitol Hill, an Army operation in the middle of the night – say two or three a.m. – could have trucked the UFO in pieces to a side door with no one around. Wrapped in canvas or sheets, the chunks of the craft would have been hauled by husky soldiers in and through double doors held open wide. Boxes of debris and the jars of alien bodies could also have been easily imported, covered with shrouds, such as blankets. All of the material would have been taken down either staircases or via the freight elevator, and lugged to the proper sub-basement storeroom under guard and set carefully on the floor or display tables. No press or public eyewitnesses were anywhere near the hush-hush process. The room would have been sealed under lock and key and perhaps armed guards, downstairs of the many floors of offices and meeting rooms, no one the wiser. That is, until we use our imaginations and information today. Now we know about their biggest secret: a damaged spacecraft, boxes of shrapnel, and three glass jars in their "dungeon."

CHAPTER SIX

Telltale FDR Memos and the Atomic Bomb

".coming to grips with the reality that our planet is not the only one harboring intelligent life in the universe."

"Seven leaked documents from three sources since 1994 provide both direct and indirect" paper trails backing MO41 events and materials, investigative author Ryan S. Wood concluded in his fascinating book *"Majic: Eyes Only,"* released in 2005. Hence it is time to take a closer look at some of the documented support for an extraterrestrial occurrence near Cape Girardeau in April of 1941 and what might have been done with it, since we've already seen the galvanizing 1947 Twining "White Hot" intel report and the electrifying 1954 Army "Special Operations Manual" on UFO recoveries. What *presidential* paperwork remains for inspection today is eye-catching, moderately revealing, and reasonably convincing.

As we all know, President Franklin Delano Roosevelt was considered an historic, outstanding, even *legendary* United States president, taking office in early 1933 and leaving it only in death, on April 12th, 1945. FDR was shrewd, manipulative, patient, and quite adept at mapping out future plans and schemes carefully and cleverly, according to those he knew and later historians. If so, it might well explain a signed and stamped memorandum issued on White House stationary, unearthed several decades after its dictation date: February 22nd, 1944. This would have been almost three full years since the MO41 incident, and reflected it in the memo's wording and content. It was unearthed by researchers in 2000, and it is very close to the proverbial "smoking gun" in this case. The credible, attention-catching Oval Office memo helps to reinforce in small ways that aliens crashed to earth in Missouri and the U.S. Army snatched up the evidence for FDR to ponder later. The tested and authenticated document also gives us clues into President Roosevelt's delicate and intricate "statecraft of space crafts," so to speak. Franklin's wise way of getting what he wanted out of others without upsetting the applecart.

The recipient of the presidential memorandum tells us a lot: Roosevelt's quietly created "Special Committee on Non-Terrestrial Science and Technology." Wow! Who knew that FDR had suddenly felt the need during his presidency to create a unique, behind-the-scenes study group that covertly scrutinized otherworldly "items" that American interests had gotten their hands on? The memo's nuggets lead to all sorts of exciting issues, but the first one that comes to mind is the most obvious: *what precisely was there to study that was "non-terrestrial" and required a special, secret ensemble of scientists?* The memo doesn't specifically say, but obviously we can guess and ask still other pertinent questions. Like, why did this unique study group not receive press coverage and Washington fanfare, book mentions later and internet exposes? What were their individual names and qualifying backgrounds? When and where did the Non-Terrestrial Committee meet? And why was Roosevelt's '44 memo to them stamped "*Double Top Secret*" (twice)? What was so explosive it had to be studied quietly by a covert committee that cagey FDR replied to by a typed, secret memorandum and not by a telephone call or visit in person?

As Ryan S. Wood succinctly put it: those in the know "came to realize the wealth of technology that lay there for

the pickings" within "the Cape Girardeau crash of 1941."
It might have become a virtual *competition* for presidential
approval of various proposed projects and financing and
glory, in developing the alien hardware into something
fresh, innovative, worthwhile, and exhilarating. Who was
going to do what, precisely, with whom, and why? And
how, and for how long, with the ET-made goodies?
"Apparently the Special Committee on Non-Terrestrial
Science and Technology had been working for some time"
on their proposals, presentations, and pitches, "to define
clear action," Ryan summed up on his revealing Majestic
Documents website. Now FDR was forced to respond to at
least *some* of what had been run up the proverbial flagpole,
but likely to great scientific disappointment he still had the
war effort foremost on his mind. Roosevelt didn't want to
go down in history as the first U.S. president to lose a war,
and this one was for all the marbles. The future of the
whole *world* was at stake. It was time to cash in all of his
chips and IOUs. What was being studied at the highest
level *had* to be converted into something useful in
American weapons systems, as we shall see.

The memo lists one of the group's leaders, Dr. Vannevar
Bush, PhD., a very brilliant, educated, and accomplished
scientist, having toiled devotedly for FDR for years in the
field of the application of scientific endeavors in warfare
technology. Much-decorated Bush was the leader of the
U.S. government-backed, Washington "Office of Scientific
Research & Development," or OSRD for short. He also
headed up FDR's National Advisory Committee on
Aeronautics, as of 1939, interestingly; this would have
covered high-tech fighter planes and early rocket research.
Lean and tall, bespectacled and married (with two sons),
masonic Van Bush (1890-1974) was well-trusted and well-
liked, although his pace was said to be generally rather
slow and academic. "No American has had greater
influence in the growth of science and technology" than

inventor/engineer Vannevar Bush, according to one Bush biographer. It was this very same leading scientist and the president's trusted physician who paired up unusually for a special closed-door session with FDR on the morning of Wednesday, April 16th, 1941, to discuss something very important. Was it about the clinical examination of the hardware (spaceship) and an autopsy of biological entities (ETs) that fell to earth near Cape Girardeau, held by George C. Marshall's U.S. Army? It was GCM himself who then met with FDR in the same office just two hours later, strictly "off the record" (see Chapter One).

Almost every matter involved herein seems able to be boiled down to what was recovered from a Missouri farm field in April of 1941; it was most definitely "non-terrestrial technology" and of a "Double Top Secret" nature. It certainly would have required an eminent scientific study group, too, perhaps stocked with trusted, experienced academic members who toiled almost daily in secretive research for the government at times anyway. And the MO41 recovery fits within the time-frame of this Roosevelt White House memorandum as it might have taken months or even *years* to get the technology from the recovered craft to be privately studied and reasonably understood by a group of patient men of science, likely assembled from the more accomplished and revered universities and skilled labs of the day, as opposed to being controlled and only occasionally scrutinized by less qualified minds within the United States Army. Thus it is safe to say that the MO41 wreckage "technology items" were the presidential memo's main - if not sole - subject matter. Thanks to events in and near Cape Girardeau, FDR had a potential secret weapon, an alien ace up his sleeve that even his war allies and perhaps some fairly close advisors didn't know about. He wanted this high trump card readied to lay on the table.

Before we dig into it further, let's remember that Mr. Roosevelt cautiously dictated this memorandum to his Oval Office secretary, who may well have been sworn to secrecy, but the subject matter (MO41) was so explosive that even *she* was not allowed to know specifics of the "non-terrestrial technology" that was being bandied about. In other words, FDR had to couch his terminology in more vague terms that only his memo's receiver would have understood, not a woman taking short-hand notes to type up later and send by courier or in the U.S. postal system. A woman who might have talked afterwards, if even *accidentally* blurting out something revealing. And if the memo fell into the wrong hands, its new owner must not be able to understand specifically what it was referring to. Espionage by various sources, foreign or domestic, that was a real problem in wartime Washington. Thus, vagueness at times was extremely important to both the sender and its recipients. Both sides of the memoranda already knew what the explosive main topic was - and now so do we.

{Note: many of the documents in this chapter can be see online at Ryan Wood's brilliant and highly recommended MajesticDocuments.com.}

The president starts out by naming two famous scientists who were involved in hush-hush "Non-Terrestrial" research projects: Princeton University's Albert Einstein and the aforementioned Dr. Bush, the remarkably brilliant scientist from M.I.T. and other respected universities, and the Carnegie Institute. The duo had obviously asked for a mysterious "separate program" to be conducted on some unspecified part of it, a serious "recommendation" for "exploratory research" as FDR called it. The Chief Executive then mentioned something of a very extraordinary nature that obviously excited him as he managed in February of 1944 the various military operations (within two main war fronts) that drove the

United States to eventual victory. Clearly FDR was in a war frame of mind. "I also agree the application of non-terrestrial know-how in atomic energy must be used in perfecting super weapons of war," to defeat America's enemies, the president states in his second sentence.

For those who might say, "Roosevelt could have been referring to meteorites and asteroid particles that had fallen to earth." The answer is definitely "no," since those things were not really "science and technology" or "new wonders" that could be applied to or as "super-weapons of war," intensely studied by brilliant scientists and academic scholars. Meteorites do not possess "atomic know how." Asteroids do not require secret studies by top academics. But certainly MO41 was so advanced that our own scientists would need to study it cautiously for three years as they struggled to carefully infuse its advanced technological makeup into our own defense systems. And as we have seen (in Chapter Two) these non-terrestrial studies apparently paid off in an alien-copied atomic propulsion device that quite possibly revolutionized American weaponry and more.

In the explosive February 1944 memo, President Roosevelt goes on to plainly inform the scientists within the special study committee that his first and highest priority was to find ways to win the worldwide war "as soon as possible." If America lost, it seems unlikely there would be any ability to further study alien objects of any kind. And Roosevelt also made clear the country had already diverted much of its research and development funding into military matters, and that this policy would continue as long as the United States remained in battles overseas. Thus, great financing of intricate, long-term secret ET studies was nearly impossible in that period. It would require asking "further support of the Treasury Department," perhaps military leadership, and most likely the congressional

Appropriations Committee for more funds, and that meant explaining to all of them just what they'd need it for, which was evidently quite unacceptable.

It should be pointed out that also in February of '44, General Marshall, Secretary of War Stimson, and Dr. Bush drove to the Capitol Building for a hushed, closed-door meeting with the Speaker of the House, the Senate Majority Leader, and the Senate Minority Leader. This was held in Speaker Rayburn's office down under the first floor. It was about a firm but secret financial request for well over a billion dollars to fund the burgeoning atom bomb project, which the three critical political leaders were told about for the first time. "The greatest secret of the war," they were told, according to historians, even kept from then-Senator Truman and his tenacious defense financing committee. This staggering sum for the mid-forties was obviously what blocked further ET research projects; winning and funding the war with the new technology (partly aided by MO41 "atomic know-how," evidently) came first.

FDR soothingly reiterated in the '44 Oval Office communique that someday he'd get behind more elaborate and better-funded explorations through "a program devoted to understanding non-terrestrial science and its technology." To reemphasize that Franklin Roosevelt was tight with OSRD's leader, the confidential memorandum mentions how the president previously "had private discussions with Dr. Bush on this subject," in addition to "the advice of several prominent scientists who believe the United States should take every advantage of such wonders that have come to us," but not until peacetime. Evidently Mr. Roosevelt was very enthusiastic about discussing MO41 implications and applications with the brightest brains the country had to offer, in strictly private conversations and settings sworn to secrecy. Likely they had exchanged ideas on the "wonders" that somehow "have come to us" for some rather special high-flown notions.

Later in the document we see President Roosevelt reference again the "proposal" by the special academic collective to study non-military applications for the real world, so to speak. Atomic energy or similar projects not infused into weapons but other items, systems, and programs, presumably. Roosevelt let them down gently, saying "I commend the committee for the organization and planning that is evident in Dr. Bush's proposal." Whatever Bush and friends were wanting out of the alien designs, it was going to be too peaceful, too pricey, and too time-consuming to produce in wartime and very difficult to assure any measure of success when completed. Thus, it was to FDR a non-starter. yet it had obviously caught his attention and imagination over the past few years and he diplomatically did not wish to throw cold water over the entire project. But what was it, precisely? We don't know, but. the original subject matter was, after all, a spaceship. Provocatively advanced, complex technology that likely sent imaginations - like FDR's - soaring into the stratosphere. The amazing MO41 ET craft could evidently transcend our laws of gravity. It apparently could transport sentient humanoid beings across the planet - but likely not very distant star systems or galaxies - at impressive speeds, beyond current manmade aerodynamic abilities. The president and his Special Committee had to wonder: could it - or a recreation - perhaps do the same someday peacefully, and safely, for human pilots and passengers?

We know the recovered, damaged non-terrestrial MO41 craft was small, and featured "child-sized" seats in its flight deck. It was so badly splintered in a key area it likely could never simply be repaired and host human pilots. But what if it were broken down, reverse-engineered, and copied, on a larger scale, using manmade materials? Then an exploratory atomic-powered spaceship to the moon, let's say, would have been conceivable to forward-thinking individuals in 1941 and the ensuing years. This secret program would take years to develop, sure, but the MO41 technology would be quietly studied and tinkered with by the finest minds in the country, and maybe, just maybe, recreated in a newer, grander form to someday take the bravest American pilots into so-called "outer space," a process that would shock and impress all the war-weary nations of the entire, troubled earth.

The memo's phrasing of it all. still makes one wonder today if FDR and some of his selected top minds were excitedly envisioning back then a kind of sustained, viable, future "space program," in addition to (and fueled by) the "atomic know-how" projects being proposed. A series of larger, reusable spaceships to explore space before returning safely to earth? After all, he just got his mitts on a real spaceship. However, this theory remains speculative, and yet backing up this galvanizing notion was none other than ally Great Britain's low-key spy ring, operating in Washington during this period in World War II.

The "British Security Coordination" espionage effort was headed by mysterious Canadian (and pro-English) spy-master William Stephenson, who was in charge of worming secrets out of American politicos and military bigwigs, intelligence and governmental figures in the 1940s, to better gauge the Yanks' war effort in conjunction with the British Empire's. What wily Stephenson found out - partly on his own, and partly through his low-key operatives - was startling indeed.

According to a contemporary author, spy Roald Dahl filed a covert report for William Stephenson within BSC headquarters in Manhattan's Rockefeller Center "in late 1944." It stated that Dahl had recently learned to his astonishment that American President Roosevelt had

strangely developed an unofficial, private "space program"!
Or at least a special *plan*. FDR had apparently been
expressing grandiose ideas in private to somehow,
someway, someday send American pilots into the cosmos,
or at least as far as the moon. Dahl had lately spent much
time with Vice President Henry Wallace, who was
disgruntled that he was not included on the '44 ticket, or
given the newly-opened Secretary of State position. Dahl
learned – likely through Wallace – that the American
president wanted to place his country's pilots on the moon
in order to plant an American flag there, to claim it as
United States territory. But how? And just where did Mr.
Roosevelt get this idea?

In 1964, noted WWII spy-turned-writer Roald Dahl
published his now-classic children's book, "*Charlie and the
Chocolate Factory.*" That is of course the popular story of
a little boy named Charlie Bucket who wins a chance to
tour inventor Willie Wonka's private candy factory and in
the end ride in his "Wonka-vator," as it was called in the
popular 1974 movie. This special glass-and-steel elevator
in the factory zips upwards at a great speed, smashes the
glass ceiling (literally), and then soars into the sky,
whereupon it zips about high over the city and countryside
as a controlled spaceship. Unlike the decade-later movie,
the book plot has the intelligently-controlled aerial craft
land at young Charlie's house, pick up his family, and fly
them away. A little-known 1972 sequel by Roald Dahl
exists, entitled "*Charlie and the Great Glass Elevator,*"
which picks up at the end of the original story, sending
Wonka, Charlie, and his family soaring into outer space,
eventually meeting up with an orbiting space station
created by the U.S. government. There they meet strange,
big-headed, large-eyed extraterrestrials, have a wild
adventure, and end up landing safely on earth, where they
are invited by chair-bound "President Lancelot R.
Gilligrass" (sounding much like "Franklin D. Roos-e-velt")

and his "Chief of the Army" to the White House. Decades ago, all of this seemed like pretty far-out science fiction for a retired espionage agent. But Dahl was no ordinary spook; he spent a great deal of the early 1940s in Washington D.C., purposely trolling the social scene, working his contacts that surrounded Roosevelt, Wallace, and intel chief William Donovan, among other luminaries, including top military and intelligence sources.

Dahl's resulting intel file on FDR's lofty space dream was evidently shelved as irrelevant and largely forgotten. until it was recalled in July of 1969. That's when the struggling but tenacious American space program finally came through for their late president. But in the minds of the public, it was for slain President *Kennedy*, not Roosevelt. It was JFK who greatly promoted the "space race" and a moon landing. U.S. pilots - "astronauts" by then - landed safely on the moon in '69 and firmly stuck "Old Glory" into lunar soil, via Neil Armstrong and his Apollo compatriots, for the world to view on live television. Then they got back in their spaceship and flew safely back to earth as heroes, eventually being invited as heroes to visit the president at the White House. Suddenly Dahl's 1944 report didn't seem so "far out;" Stephenson cabled him out of the blue to congratulate him on being right all along.

To back up Roald Dahl's secret intel claim on learning FDR's plans, we also know from two other documents - the aforementioned 1947 Nathan Twining "White Hot Estimate" and a letter from CIC/CIA leaker Thomas Cantwheel - that the U.S. Army indeed had one or more secret projects of transforming Missouri-recovered alien technology into special vehicles for advanced flights in the upper atmosphere. A new "S Craft" vehicle was created and tested in New Mexico's desert in the 1940s, perhaps while FDR was still alive, based on components of downed/discovered alien airships, Cantwheel assured in his typed confession. Could these actual ongoing confidential projects have been the cause of President Roosevelt's enthusiastic, somewhat private dream of a future reliable spacecraft for mankind's noble lunar pursuits?

Let us bear in mind that William Donovan was a longtime friend and college classmate FDR and so many other powerful men of the day. In fact, via William Stephenson's urging, Donovan was named by FDR as the first head of the "Office of Coordinator of Information," a kind of forerunner to the modern Department of Homeland Security, overseeing intelligence reports and programs from various U.S. intelligence agencies, including the military branches and J. Edgar Hoover's FBI. This critical new office was put together *in the months after mid-April of 1941*, and by July, William Donovan was named its chief. One of his best pals was Air Corps General Henry "Hap" Arnold, who might well have had his own connection to MO41 recoveries (see Chapter Four). And naturally, the two men were hardcore Freemasons.

In a fact a little lost to history but revived here, William Donovan's top OCI aide was none other than James Roosevelt, age 34, the president's dutiful Marine son and former White House confidential secretary. Wild Bill and James had authorized access to the biggest, most critical secrets within Army and Navy Intelligence, Counter Intelligence groups, and as mentioned the FBI; the duo's duties and impact became even more critical after Pearl Harbor. James was not in D.C. but readying to marry his sweetheart in Los Angeles when the circular UFO went down in April '41, but if there was any member of the Roosevelt family or inner circle who was soon entrusted with the facts about this amazing occurrence, it would have been James. He already knew plenty of other secrets, having helped his father in just about every way during the 1930s.

As we shall see in another later-leaked memorandum from the physically handicapped president, "General Donovan" was definitely one of the few to know about the Army-recovered "celestial devices" American scientists had been privately studying. And as we've seen, "Dr. Bush" was another listed by the president himself in this same memo on the explosive subject, just as Donovan was. Vannevar Bush met in the White House in October of '41 with the president and vice president regarding top secret atom bomb-related subjects, historians point out. The whole circle of connected names and collected data hangs together very tightly.

In concluding his February 1944 White House memorandum on secret extraterrestrial subject matters, prudent President Roosevelt conveyed his "appreciation for the effort and time spent" on the non-terrestrial group's proposal. The busy chief executive appreciatively congratulated the committee's ability to advance ideas to help "in coming to grips with the reality that our planet is not the only one harboring intelligent life in the universe." Here is flatly stated another very strong indication that Mr. Roosevelt was monitoring their progress fairly closely and truly believed they were all dealing with very advanced materials originating from another planet, if not galaxy or dimension. This was, of course, MO41, and by then perhaps a supposedly-recovered early '42 Los Angeles UFO and possibly even a rumored innocently-discovered, intact alien spacecraft retrieved sometime in '42 from somewhere in Louisiana (mentioned by Cantwheel in his typed confession). All of which were recovered by George Marshall's Army Intelligence officers, evidently.

It should be noted that according to records, the U.S. Army's Counter Intelligence Corps consisted of a total of 288 men by mid-February of 1941, under the command of General Marshall. In the weeks after the MO41 crash, that

CIC number not-so-strangely suddenly shot up to 513 by May 31st. This figure rose considerably even higher after the bombing of Pearl Harbor and the U.S. entrance into World War II much later that year, understandably. It appears that a program was put together to research the likelihood that not only was our country, if not our planet, being observed by an alien presence, but the frightening possibility existed in the minds of the Army officers in the know – and FDR? – that the human race was about to be attacked by "non-terrestrials" in their advanced airships. Happily this was not the case. But this was not the only increase in behind-the-scenes intelligence and scientific probes and programs related to the Cape Girardeau recoveries...

In July of 1999, a knowledgeable anonymous source sent noted UFO researcher Timothy Cooper a seven-page, typed letter summarizing all the facts (and some theories) he had gathered from his inside scoop of very quiet U.S. government intelligence studies of extraterrestrial matters. We'll call this person "Jerome," and list a startling set of reasonably believable statements he made regarding MO41 and other recovered alien hardware from the middle of the twentieth century. Fully realizing anyone can make up unproven stories and insider allegations, we must still seriously ponder what has been claimed.

First, Jerome stated he learned that sometime in the 1940s, the American government - mostly through the military, which eventually settled into the new Pentagon building in D.C. after MO41 went down - set out the funds and select personnel to carefully and covertly scrutinize recovered extraterrestrial artifacts and then try to take advantage of the unique finds in any way possible. They eventually did so in at least two main programs (that were eventually merged into one by fascinated President Harry S Truman). One hush-hush research project was entitled "Project

MAJESTIC." This small scientific study group absorbed all that was available from retrieved alien crafts to learn about finding ways to communicate with visiting, observing space travelers. Presumably, this would come about by learning their language, via the hieroglyphic-like drawings within the alien spaceships discovered, and from any possible materials or devices gathered within that dealt with their culture and linguistics. And perhaps also from studying any obvious communications gear or data found within the scout ships. Such a program would also "seek ways of detecting non-terrestrial signals," Jerome claimed. This all seems logical enough, the worthy (or "majestic") approach apparently being: *"let's talk."* The American government obviously wanted to find a way to open a dialogue and privately communicate in peace with their rather aloof but advanced visitors.

The second program Jerome discovered was called "Project JEHOVAH," and this evidently was designed to "back-engineer the hardware" left on planet earth by alien astronauts. JEHOVAH was also to "research the physics" involved in the propulsion mechanisms and atomic engines and technical systems of the found ships, so that perhaps American scientists and engineers could then reproduce their own model. This then was most likely what was turned over to Dr. Bush to guide, with his Office of Scientific Research & Development, headquartered at 1530 P Street in Washington D.C., along with some guidance of the atom bomb program, or "The Manhattan Project." It was most probable that FDR's 1944 memorandum was addressing this specific effort, in order to turn the golden understanding of the "atomic secrets" into useable weaponry, or at least very advanced vehicles for delivering "ordinary manmade" weapons. This would presumably not be what peaceful non-terrestrials would have desired, to say the least. Such a program would also likely have been responsible for producing the know-how to create a reusable space vehicle for American pilots to take to the moon and beyond, as we now understand.

Both of the highly classified MAJESTIC and JEHOVAH groups were eventually overseen by a special, secretive twelve-man panel that was aptly dubbed "MJ-12," Jerome explained to researcher Tim Cooper. These one-dozen powerful individuals hailed from "military, intelligence, and scientific institutions." United States presidents from Truman to Nixon were supposedly briefed at times on the progress of these covert study programs, and still other smaller or newer ones under the MJ-12 umbrella, until in 1969 Mr. Nixon supposedly disbanded the entire governmental operation, turning a few of the remaining programs over to private institutions and corporate research labs.

The entire February 1944 FDR memorandum to Dr. Bush and his alien-themed Special Committee brings us back to a similar memo mentioned earlier. One that Franklin Roosevelt dictated and sent from the Oval Office of the White House, uncovered by researchers in 2000 and authenticated after great scrutiny. It was also dated in February, this time in 1942, about ten months after MO41 took place. It was also stamped "Top Secret." It was undoubtedly sent by special military courier across town, to the old Munitions Building where the War Department was holed up, featuring offices for the Army Chief of Staff. In this remarkable document, FDR tells the recipient, General Marshall, "I have considered the disposition of the material in possession of the Army that may be of great significance toward the development of a super weapon of war." So here we notice that once again, busy F. D. Roosevelt was chiefly worried about the ongoing global conflict and how to gain the upper hand via something unnamed yet potentially exciting that the nation had mysteriously acquired. Hmmm, what could it be?

The president's specific memo phrasing again is eye-catching: material that could make for a "super weapon." Something that General Marshall's army had been controlling and generally keeping out of the sustained reach of trusted top scientists like Van Bush until early 1942. Roosevelt urged "finding practical uses for the atomic secrets learned" from the tantalizing technology. This required close, prolonged laboratory scrutiny and testing. FDR would meet with GCM once in a while, mostly at the White House Oval Office, and sometimes chatted with his Army Chief of Staff over the telephone, yet the president felt compelled to send the preoccupied, highly decorated general his thoughts on paper and make very clear it was now time to free up the "celestial devices" for the good of the fledgling war effort. MO41 components, it can be very reasonably assumed.

Again we see that the otherworldly materials couldn't be named openly in the '42 memorandum as this would have been a national security breach and fodder for gossip by aides, secretaries, and couriers in typing and sending/receiving the document. Or for wartime spies, potentially. And once again, Dr. Vannevar Bush is mentioned (three times), with FDR giving Marshall his permission for this imminent scientist and others to "proceed with the project without further delay." Obviously there had been a hold-up in the extended close scrutiny of the unique MO41 materials from their sudden arrival in April of '41 to the need for an official presidential memo on the matter in February of '42. Of course, in that time America's armed forces and the scientific community had galvanized for war after the Japanese attack on Pearl Harbor in December '41, turning everyone towards the probability of a long slog for the nation just to survive and help other free and not-so-free countries do the same, let alone win.

At the noteworthy '42 memo's conclusion, Roosevelt authorized that Marshall could "speak to me about this if the above is not wholly clear." If he could not reach FDR, then war-busy GCM was specifically instructed to keep all conversations about the amazing "new wonder" restricted to "General Donovan" - tellingly the first man listed - and "Dr. Bush," as we have seen, plus Henry Stimson, the Secretary of War. Pretty elite company in 1942. In fact, the most powerful and reliable men in the country, perhaps the world. If MO41 were to be entrusted to anyone in the nation, they were the logical ones for Roosevelt to have picked to handle it, at least during wartime.

The '42 White House memo to Marshall appears to neatly set up the '44 memorandum; the latter being a response by FDR to Van Bush and his Special Committee's requests in desiring to even more thoroughly study and exploit MO41. The main overall message of this early '42 memorandum is clearly summed up and spelled out by the president; it was Roosevelt basically telling Marshall: "Quit hoarding the Missouri spaceship and let our top academic, non-military brains at it more extensively, since your army and limited scientific access aren't getting us anywhere." One can logically conclude that GCM was apparently doing something at times with the ET artifacts that kept the trusted high-level scientific world at bay for much of the remainder of 1941 and early '42.

Winning the ongoing worldwide war. Creating a reliable atom bomb that could help wipe out the German and Japanese aggressors in World War II. This was evidently for FDR the *main* purpose of intensely scrutinizing the MO41 recovery, even by non-military scientists. The president urged this top secret matter be discussed only by a chosen few, named them, and then concluded with his usual emphasis on first "the daunting and perilous" conflict. Once victorious, Roosevelt calmly assured

Marshall, "the Army will have the fruits of research in exploring further application of this new wonder." It's difficult to conceive of a single alternate source of unique matching material in this particular early '42 memo's description that wasn't from MO41 alien hardware. Mr. Roosevelt clearly wanted Dr. Bush to get this unique study program rolling with greater expediency, although that simply wasn't the normal way the slow-moving, ponderous Van Bush operated.

Another once-classified memorandum, this time sent from Dr. Vannevar Bush to the White House, backs up the notion that FDR was eager to push the weaponization of the "non-terrestrial wonders." This Bush memo was also leaked in 2000 but was originally dictated and sent just a day or two after top secret news arrived regarding the recent early-July 1947 New Mexico UFO crash discoveries. In it, Dr. Bush patiently explained in this frontispiece letter to President Harry Truman how then-President Roosevelt wrote to him back in April of 1944 - MO41's third anniversary - and asked Bush four important questions regarding "non-terrestrial sciences." {This specific FDR letter to Dr. Bush has not been unearthed.} Over two years after Roosevelt's sudden death and after plenty of scientific studying of MO41 and perhaps other recovered crafts, Bush informed Truman that Roosevelt wanted to know all about not just weaponization of ET technology in time for use in the continuing world war, but also general scientific uses for after the dreadful conflict ended and then for many years into the future; i.e., FDR summarily asked: *"What can the government do now and in the future to aid alien crash research activities?"* Good question! What was the answer? "Progress in other fields is important, such as material fabrication" {reverse-engineering, in other words}. Van explained further, "but the program for non-terrestrial science in my report warrants immediate attention." Despite all of the behind-the-scenes findings, fussing, and fears over the '47 Roswell incident, Dr. Bush insisted on presidential attention to his unnamed pet projects first, evidently born out of MO41. He cleverly included for Harry Truman with his memorandum an attached "plan I am proposing" which was supposedly acceptable to the low-key "Non-terrestrial Science and Technology" group that F. D. Roosevelt set up in '42, a strategy that would create a "single mechanism for implementing the recommendations" by the secret UFO study group ("MJ-12," most likely). The Bush plan in tangible letter form has not yet been discovered. Obviously by mid-1947 World War II was over but the need for understanding and main-streaming the scientific secrets - like a special copied alien propulsion system device - learned from the retrieved otherworldly hardware was still very important and needed to be supported by the current chief executive and future administrations, Bush felt. He did just that; President Truman simply signed his name and "OK" at the bottom of the copy researchers discovered and scrutinized.

It wasn't just hardware the late FDR wanted to know about, Vannevar wrote to President Truman, but also "non-terrestrial sciences, including biology and physics." In other words, the makeup of the ET pilots - their brain matter and neurological systems - and their physiological relationship to the technical wonders they controlled within their spaceship, plus the atom-powered propulsion system's calculated formulas within it all. And evidently out of that latter study came possible improvements within The Manhattan Project, just in time to win the war as the late president had hoped and dreamed.

It should be pointed out another fact that becomes obvious in reviewing the uncovered FDR-to-GCM memos: studies of the inner workings of the MO41 craft resulted in a preliminary report that declared that the splitting of the atom was energy harnessed as a power source by/for ETs.

In other words, for Roosevelt to speak in the memorandum about the Missouri-retrieved alien ship's "atomic secrets" means the president had to have been previously briefed (by Dr. Bush, most likely) on the general atomic makeup of the crashed vessel's engines and/or propulsion system, sometime in the latter half of 1941 or early in '42. {Quite possibly this was done at a classified White House meeting with President Roosevelt and Vice President Henry Wallace, logged as taking place on October 9th, 1941. At this confab, FDR immediately approved Van Bush's OSRD to develop the atom bomb, and as we can see through the leaked memo afterwards issued a direct order to keep this secret project limited "to Secretary Stimson, General Marshall," Dr. Bush, and also Dr. James Conant, Bush's close academic compatriot.} These MO41 component studies had been fascinating, complex, and ongoing, as they logically would have been, but apparently unfettered and sustained scientific access to the salvaged ET materials for scientists to this point was an ongoing sticking point. Bush must have complained to the president that this was a substantial problem in developing anything useful from the heart-pounding recovered extraterrestrial technology.

Perhaps it is again mere happenstance, but the complex atomic energy research project undertaken by scientists and teachers at some of America's leading universities and think-tanks had actually begun in the late 1930s, fueled by the discovery of ways to harness the element uranium to create enormously powerful explosives. The atom bomb, of course. Even world-famous physicist Albert Einstein lent his name to an academic letter to influence President Roosevelt back in 1939 on this volatile matter, describing the many dangerous but advantageous possibilities of this sensitive project. The issue of splitting the atom to derive great power had been in reality batted about by scientists for two decades. Yet few today recall that it wasn't until February of 1941 when American scientists toiling in

Berkeley, California, discovered the element of plutonium – soon named after the planet Pluto, for some reason - that the U.S. nuclear weapons program accelerated at a much greater, more definable pace. Scientists were seemingly at last able to put the key ingredients together that would lead to a terrifying nuclear detonation device. Much excitement followed in the next several weeks, within the academic and scientific community, about the possibilities from the plutonium discovery and what it could mean for American's future energy and arms industries. But it wasn't quite "there" yet. Something critical was missing. Thus the great elements of the atom bomb were still being pieced together when aliens abruptly smashed into that obscure Missouri farm field in the Spring of '41 and left their high-tech machinery (and bodies) behind. And it certainly has been suggested in many UFO/ET books, web sites, and publications that the advent of the nuclear age is precisely what has drawn extraterrestrial races to steadily and stealthily visit earth to observe humanity in the first place. Or at least to accelerate their visits, which may have been going on for centuries before, or even eons. Extraterrestrial races have supposedly been warily monitor mankind's "control" of this exciting, new, but insecure atomic power, in this hypothetical theory. Advanced aliens are supposedly worried we're going to recklessly blow ourselves up, wipe out all species on the planet, and pollute space with harmful radiation, at least that's the theory in a nutshell. Could this be why three members of an advanced traveling race of "Grays" were tooling around 1941 mid-America in their scout craft at that point in time?

Of course, history records that the United States was the only nation to deploy and detonate a nuclear bomb on an opponent in a conflict, taking many thousands of lives with the mind-numbingly destructive August 1945 explosions at Hiroshima and Nagasaki in Japan. There was some talk at that time of dropping an atom bomb on the capitol city of

Tokyo, should the bellicose and defiant Japanese military and their manipulated, isolated emperor not surrender in what became the waning days of World War II. Back in early 1941, the Japan's nationalistic leaders were quietly at work on their own atomic weapons program, as well as dastardly secret plans to undertake a risky sneak attack on America's military might, and this was eventually carried out of course at Hawaii's Pearl Harbor on December 7th. However, in the last few years, it has been discovered that America's military elite had also developed a secret extreme contingency plan to do the very same thing to Japan, if it was deemed absolutely necessary. To abruptly bomb with no warning a number of their assets in a desperate sneak attack, if it was thought possible to head off a possible all-out war and to weaken the resolve and technology of their potential enemy. All this was building to quite a boil as the rest of 1941 steamed forward. Which nation would strike first? Which nation would develop the frightening atom bomb first? Which nation would prevail? It was all up in the air - and so were extraterrestrial observers, literally, it seems apparent.

It's interesting also to note some other pertinent facts. For instance, Cape Girardeau is situated roughly at the east-center of the North American landmass, not quite geographically and population-wise but still fairly close. It was peaceful and obscure, safe and typical of American life during a time of growing world war on other continents, and even within the earth's skies and oceans. Alien entities not wishing to get shot at might have chosen an orbit, or a flight path, over this general area when visiting the planet. When inspecting a globe and tracing the latitude line from the city of Cape Girardeau (around 37 degrees North) one sees it travels straight back across the planet to Tokyo (or close to it) and very near the two atom bomb-destroyed cities. Wildly, a factory next door to the masonic lodge in Cape Girardeau would go on to play an important part in developing and producing a small but key component within the atomic bomb, and still other weaponry, communications, and transportation technology to help defeat the Japanese empire and win WWII. Several citizens of Cape Girardeau - including its mayor and chief of police back in April '41 - fought the Japanese military in the Pacific theater. Even more remarkably, a United States ship called *"The S. S. Cape Girardeau"* aided in this war effort. To bring things around full circle, Cape Girardeau was also a town known well to the very man who authorized the dropping of FDR's long-planned atomic bombs on Japan, President Harry Truman, of Missouri. As a U.S. Army enlistee in the humid summer of 1906, *Harry actually lived in Cape Girardeau* and trained at the city's armory and on the nearby Mississippi River. Truman then later visited the city a few times when campaigning for state office, even later when retired and long out of government. {Even FDR once campaigned in obscure Cape Girardeau, as a vice presidential candidate in 1920.} Additionally, as president Harry was once famously quoted as opining publicly: "I can assure you that flying saucers, given they exist, are not constructed by any power on earth." He also admitted on camera to reporters in 1952 that as president he discussed "UFOs" with his military aides "at every conference we had," which is equally stunning. Likely in '41 Harry found out about the alien crash and later a very ill Franklin remarkably added him to the '44 ticket above V.P. Henry Wallace and all other possible running mate candidates, tellingly. FDR's closest pal Harry Hopkins told insiders the president "wanted Truman all along." Was MO41 the influencing factor, one that secretly changed history in so many ways?

Let us bear in mind that William Donovan was a longtime friend (and college classmate) of FDR, and to so many other powerful men of the day. At this point we must ponder the telling words of Otto Krause, a nuclear physicist

who was imported from Germany following World War II
in "Operation Paperclip." Krause was put to good use in
various American nuclear research sites, including at
Livermore Laboratory from 1962 to 1967. And what is he
known best for today? Having told at least one man, a
government-hired photographic expert, while at a Nevada
nuclear test site in 1962 that the American scientific
community completed the atom bomb thanks to very
advanced alien technology culled from a crashed "disc"
that was hauled from a landing site in Missouri "in 1941"
to secret examination labs. And that this governmental
hushed alien know-how assimilation project was
considered "more Top Secret than the atom bomb itself."
Thus at last we have a more firm understanding of a rather
famous, eye-opening statement made by another German
scientist put to work for the U.S. It seems that a 1954 press
conference, Professor Herman Oberth – considered by
some to be the father of American rocket science and the
new "space age" – was asked a question about how our
technology was suddenly so advanced. "We cannot take
credit for our record advancement in scientific fields alone.
We have been helped," Oberth replied candidly. When
asked by *whom*, the repatriated ex-Nazi replied, shocking
his audience: "The peoples of other worlds." Oberth went
on to speak publicly about "flying saucers," saying they
"are real" and "from another solar system." He knowingly
added that such vehicles were "manned by intelligent
observers who are members of a race that may have been
investigating earth for centuries." He also strikingly stated
for the record that such alien spaceships were "conceived
and directed by intelligent beings of a very high order. And
they are propelled by distorting the gravitational field,
converting gravity into useable energy." These highly
informed remarks seemed off-the-wall to newspaper
readers in the 1950s, but are far more fathomable and
sensible now in light of MO41's copied atomic technology.

In returning to the subject of the tell-tale paper trail on MO41, there is no evidence to date that Franklin Roosevelt ever shared the Cape Girardeau discovery with either the friendly British (and his masonic friend Churchill) nor the enigmatic Russians (led by inscrutable Joseph Stalin). By early 1942 the Soviet Union was a new ally in the fight against treacherous and ruthless Japan, Germany, and Italy in WWII. In the February '42 Oval Office memo FDR carefully reiterated his decision not to discuss the classified matter with the Russian government. "I disagree with the argument that such information should be shared with our ally the Soviet Union," the Chief Executive dictated to Marshall in February of '42. This tells us that there had been a significant ongoing discussion of MO41 within FDR's inner circle, with someone - undoubtedly Roosevelt's consistently pro-Russian advisor Harry Hopkins - putting forth the idea that Soviet leadership and Communist academia might be able to help Americans utilize the alien technology secrets for shared future weapons of war, and still other uses, such as energy systems and lightweight battle hardware.

There has been some controversy since the 1990s as to whether FDR's close friend and aide Harry "The Hop" Hopkins was in fact a secret Soviet informant and willing agent. Fiftyish, frail Harry's official title was "Special Assistant to the President," becoming Franklin virtual shadow by the late 1930s, accompanying him in his personal and professional life. He often sat in on Oval Office conferences, right next to Franklin. So close to Roosevelt was Hopkins that he actually lived in the White House with his boss, working or studying in the famous mansion very near the First Family. *Records show Hopkins was with FDR on the night of April 12th, 1941, and by the president's side at top level Roosevelt Oval Office meetings with George Marshall (among other big-shots) April 15th and 16th*. Thin, often-unhealthy Harry *had* to have learned plenty, either directly or through overhearing administrative and/or Roosevelt family conversations and gossip. "The Hop" was a trusted (by FDR) key player in strategy of all kinds - including preparations for possible war and the atom bomb - in the Spring of 1941, but was frankly not all that trusted or liked by other members of the administration, according to historical researchers. The debate by scholars, historians, and investigative authors for the past three decades on Hopkins' true inner loyalties and politics has been fascinating, but somewhat inconclusive. Was unassuming Mr. Hopkins truly at heart a Communist? And a Red spy? A treasonous double agent, selling out American interests? Either way, Mr. Hopkins made a rather sudden and urgent call on Russian leader Stalin in the months after MO41. The Hop suspiciously met alone in Moscow with "Uncle Joe" in private (with only an interpreter) to relay some very top secret information. Harry sure had one heckuva a scoop to relay in the incredible Cape Girardeau recovery, should Hopkins have indeed been a traitor, eager to spill atomic and/or state secrets.

Yet another later-leaked, bombshell memorandum - this time sent by General Marshall to President Roosevelt, dated March 5th, 1942 - urged the creation of a special new behind-the-scenes group that would study public or military sightings of unidentified flying or landed objects that came into American airspace. This classified private memo was likely a direct response to not just MO41 but also a strange incident that occurred days earlier in southern California. That's when the anxious U.S. Army opened fire on unknown aircraft buzzing over L.A., supposedly downing two unusual, if not alien, airborne vehicles, which were very quietly recovered. What is sometimes referred to as "The Battle of Los Angeles." But Marshall's March memo was also a direct response to FDR's request on February 27th, '42, to allow closer scientific study of the MO41 materials by Dr. Bush and his special group. Marshall's memo was as if to say to FDR: *"Oh yeah? If brainiac Bush and his secret team get the Missouri prize my boys recovered, then I get to form my own secret team to handle all the future alien recoveries on my own."*

Both the top secret FDR '42 and '44 memorandums passed through General Marshall's Office of the Chief of Staff (OCS), which was located at the War Department building in Washington, on 20th Street and Constitution Avenue, just down the street from both the White House and the Capitol Building. It was here they and other classified communiques were carefully read in private and pondered. War was the constant state of mind, if not affairs, for both Marshall and Roosevelt from 1942 to 1945, plus their staffs at the time, and spies were probably suspected here and there. Communications had to have "Top Secret" or higher stamps, and phone calls discussing high-level issues were often kept to a bare minimum. Outside of those who helped recover the evidence, George Marshall likely discussed the explosive nature of the MO41 crash and its remaining artifacts with basically only two men: conservative Secretary of War Henry Stimson (in his mid-seventies), and the liberal president, both men plagued with health problems and largely desk-bound. It is highly probable that what little was ever put on paper about the hushed incident and the U.S. Army's possession of its recovered fruits was destroyed over the years, before or after Marshall's death in 1959. But a few leaked documents - carefully proven of their authenticity and subject matter by Ryan Wood and his scholarly father, Dr. Robert Wood - have survived, thanks to secretive supporters of UFO research, but even these documents don't specifically refer to the crash event or alien hardware in direct, descriptive terms.

Apparently Marshall's March 1942 memo proposal was approved readily by FDR, for GCM and his crew to handle personally current and future "non-terrestrial" craft recoveries. The unique new intel unit that came out of this suggestion was taken directly from the Army's Counter Intelligence Corps, entrusted to only a few elite officers George Marshall new and approved there, and given the title of the "Interplanetary Phenomena Unit." {The CIC's IPU for FDR run by GCM on ETs and UFOs was created ASAP, in other words!} Other recently-uncovered, sensitive government documents referred to this unique new 1940s military ET group now and then. No one seems to have stepped forward over the decades to deny that the Interplanetary Phenomena Unit was genuine. General Marshall ran the IPU operation in a tightly-controlled environment, using top Army intelligence officers to discreetly investigate strange aerial sightings and other crash sites, all without giving the media or the public much of a clue what was truly going on. Marshall had little interest or trust in snoopy J. Edgar Hoover's FBI; the Secret Service; Military Policemen; or any other investigative body. Possibly George Marshall was so piqued by the mysterious otherworldly subject he wanted to dig into it all for his own curiosity, utilizing his own top, trusted soldiers as detectives, bringing him reports and perhaps even more crashed materials.

As we've seen, Dr. Bush seemed frustrated in early '42 with his lack of sustained access to the MO41 materials. Where were they being kept by George Marshall's armed forces, precisely, after the apparent Capitol Building novelty wore off? What was GCM doing with his Army prize? A fascinating tidbit has emerged from a pair of UFO researchers regarding the claims of a talkative woman who said her son - Guy Simeone - was once a part of a massive military training program in North and South Carolina in October of 1941...

According to writer/researcher Leonard Stringfield, while at these huge Army-controlled "war maneuvers," Guy "had been involved in an exercise" to recover a crashed UFO. Where exactly it crashed originally was not stated, but if the autumnal recovery process was a mere "exercise," as Mrs. Simeone was phrased, was it a genuine accident on Carolina soil? Or was it the breathtaking MO41 materials from the previous April, strategically planted for a topnotch army war games intelligence crew to find, crate, and hush up? Was it important to taciturn General Marshall to gauge his top men's psychological reactions and procedural training abilities to alien commodities?

Ponder the description of what the Mrs. Simeone said her son Guy talked to her about, in private, at the time: the discovery was of a rounded, silvery spaceship that had somehow cracked open. A cockpit "command module," or "control room" was visible inside the metallic ship. It featured a few small seats and strange interior markings that he could not make out. And there were also some little dead bodies "from space," three or four odd beings with big heads that featured "large black eyes, like a bug." Precise matches for what Reverend William Huffman said he saw, six months earlier near Cape Girardeau, Missouri.

Yes, this Carolina "recovery operation" sounds eerily like the MO41 find; both were controlled by the U.S. Army, especially when the all-business Marshall was around, evidently. It so happens the Carolina war games were planned partly by and for General Marshall, who craved great realism in battle rehearsal strategies, maneuvers, and execution when it came to his troops. Marshall was so hands-on he visited the simulated Dixie events and continued to monitor their progress when he was away. He and the army brass employed over a half-million troops in the proceedings, featuring two armored tank divisions, live ammo, and the U.S. Army Air Corps flying overhead, dropping men and materials from the skies. This remarkably realistic battle practice was conducted from October 6th to November 30th, 1941. Foresightful GCM felt sure a dreadful war was coming someday for U.S. forces, and indeed just *days* after the last men and vehicles were transported home from these Dixie war games the Japanese military undertook their ruthless Pearl Harbor assault, triggering America's entry into WWII.

Was General George C. Marshall really so deadly serious about training his top soldiers that he would import the shocking MO41 materials still under his control, to recreate a UFO crash on the ground for training purposes? Recall the early 1942 White House memo, where Dr. Bush evidently had been complaining often about the army's possession of the crash items and his lack of unfettered access to them for proper scientific study, FDR instructed GCM to cooperate with Van Bush and allow his research team closer, sustained scrutiny of it all. This telling presidential directive now makes more sense if the general was indeed out using the MO41 goods on the ground in Carolina and perhaps elsewhere, to carefully program his topnotch Army crews on how to box up and cover up the amazing "discoveries" out in the woods. And how to sneak the amazing "recoveries" to "a nearby Army post" where they were covertly "stashed" during the "games," according to the mother's testimony of what Guy brazenly informed her. All to better prepare the military elite for possible further incursions - or even invasions? - by non-terrestrial entities.

The damaged Carolina alien craft - we'll call "CAR41" - was alleged to have been about ten feet high by fifteen feet wide. If it truly crash-landed in the region claimed, then we're supposed to believe it "just happened" to plummet from the sky where the U.S. Army and occasionally General Marshall himself were already situated and "just happened" to find it. This seems incredibly coincidental and pretty darn unbelievable. It all took place some months after MO41 was allegedly stored down under the U.S. Capitol Building, where frankly it did no one any good after the initial excitement wore off. If nothing else, CAR41 would have been critical to gauge the average soldier's psychological and physical reaction to the other-planetary items, and assess local reaction to any rumors that might have leaked out at the time, in addition to the military's time and efforts required in secretly smuggling the goods to a safe lock-down location for secure study.

The whole startling notion for MO41 manipulation in Carolina seems very plausible, respected researcher Ryan S. Wood remarked in 2014. It would mean that the MO41 airship and well-preserved dead bodies were quietly flown or trucked into the Carolina Maneuvers and carefully placed on the ground by trusted intelligence officers for a secluded bunch of soldiers to find and learn how to properly handle, perhaps all personally quietly overseen at a distance by the serious Army Chief of Staff. Possibly the men involved were only told that it had crashed there and did not realize - at least at first - that this accident event was *staged* and actually took place somewhere else. Certainly it appears that the CAR41 situation was not taken too seriously by one young man named Guy as he told his mother all about it. If it was "for real," then Mrs. Simeone might never have been briefed by her likely sworn-to-secrecy son, and she would then not have later tipped off UFO investigators like Leonard Stringfield. So all signs seem to point to a rigged yet realistic CAR41 recovery "exercise" based on either cleverly recreated MO41 artifacts, or the real deal.

Ten years after the April 1941 aerial accident near Cape Girardeau, an esteemed scientist, Dr. Rolf Alexander, met with George Marshall in Mexico City. The highly educated

man of biophysical science and several colleagues, plus
still other people at the Mexico City airport - while waiting
for Marshall to arrive - had witnessed a "flying saucer," or
advanced alien spacecraft, buzzing over the runways and
terminal for a while. Many in the crowd had whipped out
cameras and home movie equipment to record the amazing
sight, but both Mexican and American authorities swiftly
moved in afterwards and confiscated their photographic
evidence, although the story still made the next day's
Mexican newspapers. In private, Rolf met with George and
they discussed the enthralling UFO sighting, then the
enticing topic of alien visitation in general. The respected
biophysicist later recalled that the general (and former U.S.
Secretary of State for President Harry Truman by 1951)
seemed quite well informed on the subject, and told Rolf
the low-down, away from the prying eyes and ears of the
press. What the worldly general knew and discussed likely
made the scientist's eyes bulge wide.

Extraterrestrial crafts and the bodies of their alien
occupants had been recovered in three different crash
episodes on American soil as of 1951, Marshall explained
to Alexander in their Mexico City huddle. On each of these
occasions, the landing attempt ended in disaster, with the
vehicles cracking or exploding open, disgorging the non-
terrestrial crew, who were then sadly unable to breathe
Earth's atmosphere. Thus the other-planetary visitors
"were literally incinerated from within." This seems a
strange remark, unless it t only for the aliens' respiratory
systems and perhaps a few other inner bodily organs being
"burned to a crisp," simply finding our planet's richly
oxygenated air inadequate, if not toxic, for their biophysical
needs. Thus it wasn't the crash itself that killed the aliens,
it was their exposure to earth's atmosphere, once outside
their broken craft that did them in. Our oxygen levels were
simply inappropriate for their respiratory systems. This fits
in perfectly with MO41 information via Reverend William

G. Huffman's recollections, through his wife and granddaughter, Charlette Huffman Mann (see Chapters One and Two). The dead Cape Girardeau space visitors had no visible injuries at all.

Further, humorless Marshall somberly explained to Dr. Alexander, it was the private opinion of American military and scientific researchers that the aliens were generally friendly, perhaps peaceful scientists on missions of high-minded studies of human and animal existence, and posed no real threat to mankind. Marshall also tipped to Alexander that the U.S. government took the tiny minority public reaction - over-stressed by the media - to the Orson Welles radio broadcast *"The War of the Worlds"* quite seriously. The American government, the old Army man said, did not want that late October 1938 situation of curiosity, fear, and even panic (by very few, actually) repeated on a much larger, even nationwide scale, should the fact that we are being visited and observed by unusual-looking aliens, no matter how peaceable, come out to the general populace. It might be too disruptive somehow to the common way of life for Americans. Thus, great secrecy and convincing cover stories were put out by the U.S. military and certain governmental intelligence operatives to shade the truth, in order to protect the populace from extreme reactions and the disruption of American society. It was their duty to cover up and lie, for the good of the country, in other words. This also helped to keep the real story away from the prying, exploitive Soviet government, full of scheming communists who would undoubtedly leap on the otherworldly facts and figure out how to turn it into a propaganda war weapon, or even trigger a mass panic of some kind with a rigged-up "alien invasion" situation in the United States.

In retrospect, it seems President Roosevelt and President Truman were simply trying their best to keep all aspects of the regular American manner of day-to-day living humming along without disruption, and their trusted friend George Marshall was quite willing to go along with this somber, noble, secretive plan. In keeping the lid tightly on, they felt themselves to be loyal and wise Americans, good stewards of our country's governing bodies. After all, if the people learned our planet, our country, was being visited by creepy little gray aliens who can seemingly defy our laws of gravity in dazzling advanced spaceships, would many citizens truly panic? Would there be genuine hysteria, not like the media-driven exaggerations from October of 1938? Would some give up their faith in science? In their religion? Their trust in the military to adequately protect them? Would some even quit their jobs or leave home in search of alien-theme cosmic answers, or imagined future ET contact? The stock market, national economy, factory production, school attendance, military enlistment, dependence on faith. these all and more would possibly be damaged, perhaps beyond repair. Or at least, that was the way of thinking by the fear-based decision makers in Washington. All because a few people - mostly in New Jersey - called up police stations and radio stations after foolishly or innocently misinterpreting a staged radio play back in '38.

American authorities had possession of three crashed extraterrestrial spaceships by 1951, General Marshall supposedly asserted. Unfortunately, Dr. Alexander had nothing on tape or a paper trail to deliver to his scientific colleagues or the media on the stunning Marshall explanation. Rolf had only his good word, and to many, that simply wasn't enough, yet what Marshall supposedly imparted that day in Mexico City certainly sounds perfectly in line with the Missouri crash facts to a T, or an *E.T.* It also explains why the general and the United States government would keep a tight lid on the story, right from the start then and to this day. Dr. Alexander could only move on to his latest projects in biochemistry, and revealed the story to a UFO researcher only many years later, his sound reputation still intact.

Another White House document issued by President Franklin Roosevelt is nearly as important to consider as the top secret ones mentioned herein. In the days and weeks after the events of mid-April, 1941, FDR was pondering how in the world he would be able to protect American citizens from any sort of aerial attack. That spring, conquering military dictatorships like Nazi Germany and Imperialist Japan had not actually threatened to harm the United States. Yet the handicapped president was consulting his aides on possible aerial assaults on the homeland and came up with a protective idea, one that resulted in an official Executive Order, Number 8757, finally issued on May 20th. This intriguing act created a federal emergency agency entitled "the Office of Civilian Defense." *OCD was launched specifically to coordinate various groups of Americans in efforts to watch the skies for any suspicious foreign aircraft!* Air raid wardens and aircraft spotters, blackout coordinators and even fire brigades were assembled in most communities within U.S. states in the weeks and months that followed the executive order from the Oval Office. Sighting unusual, foreign airships or sudden crashes and subsequent fires... FDR was obviously deeply concerned about these subjects after MO41. Coincidence? On top of all of this, the Civilian Air Patrol was also initiated from this mid-May presidential order, to assist military pilots in keeping a closer eye on American airspace. Even of further suspicion and interest is the fact that even to this day many of the records on the formation of this Roosevelt act are still considered "security classified" and off-limits within the National Archives in Washington D.C. Certainly, if anyone questioned the sudden obsession by President Roosevelt on his urgent need for citizens to watch the skies for strange aircraft, he could simply point to the bombardments from above in the ongoing merciless war efforts in other countries, far away. But now we know he had something else on his mind that would fit under this protective umbrella. Something he just couldn't bring himself to spell out to the American people, but troubled him greatly, although he did possibly telegraph his fears in a national radio broadcast on May 27th (see Introduction). One person hovering around Roosevelt for this important White House address to the world was none other than William J. Donovan, soon to be officially named the new intelligence office chief and as mentioned previously, listed as someone in the know on recovered top secret "celestial devices."

At this point it is certainly valid to ask: is it really believable that President Roosevelt could maneuver and manipulate the entire alien situation with such great secrecy, keeping the shocking blockbuster MO41 truth from the press and the public in 1941? Well, consider two important facts about FDR's presidency. First, he contracted polio in the 1920s, and was never able to fully walk on his own again. Franklin and his dutiful aides kept the press from telling - or showing through photographs and/or newsreel footage - the full startling truth. The public knew he had suffered some "infantile paralysis" but understood he was otherwise fine and could walk and move about comfortably anyway. Americans did not realize FDR was often physically weak and almost entirely confined to a wheelchair, or for public occasions the shaky use of a cane and heavy metal leg braces when attempting to stand. He could only "walk" awkwardly and briefly on stiff legs within those braces, in some amount of pain, and always with someone's help. Most Americans did not find out that Roosevelt was a considerably disabled and diseased president until well after he died, and were rather shocked upon learning the full truth. By at least early 1944, he was diagnosed with various serious maladies, like congestive heart problems, and grew considerably sicker. Admiral Ross McIntyre, the president's physician, fibbed to the press and simply stated FDR merely had a "touch of the flu." Ross lied so many times to people about the true nature of Franklin's health, his closest aides got fed up. They set up an appointment with another, more qualified physician, who discovered to his horror Roosevelt's heart was greatly enlarged and he had still other serious health problems as 1944 wore on. All involved were quietly pondering how much longer Roosevelt had to live, including the chief executive himself.

Tellingly for the MO41 story, President Roosevelt instructed his obedient Secret Service guards to abscond with the film of any camera-holder who tried to record his difficult attempts to walk to or from public speaking engagements. Newsreel cameramen and still photographers of the day always shut off their equipment dutifully, and only rarely did a newsman or local shutterbug have to be bullied into giving up their cameras to have the film roll removed and destroyed when they balked at this general order. It was obviously a very different era of press cooperation and public obedience. {Cover-ups actually grew worse as 1941 wore on. After the Pearl Harbor attack later in December, the "United States Office of Censorship" was established and embraced, helping make sure that only approved stories made the news. The Roosevelt administration had great sway over this government effort, tightly controlling and managing news "for the good of the country" and its war effort that few recall in this media tell-all age today.}

Secondly, paralysis-plagued, increasingly ill FDR had at least one known mistress, perhaps more, despite his marriage to Eleanor and public image of family morality and loyalty. He had long since ceased intimacy with his famous spouse and the two lead largely separate lives with separate bedrooms, even separate homes. Franklin privately enjoyed the company of fairly attractive and intelligent women who were loving and caring (platonically) in keeping his troubling secrets in return, but he also carried on a sexual affair at times with the adoring Lucy Mercer. Yet this startling fact was also held quite in close confidence by the knowing press and by FDR's family and friends, administration and reporters, until decades after the great man's demise. Just weeks after MO41 occurred, FDR supposedly told a friend about his adroit ability to cover his many secrets and schemes with compartmentalization and deception: "I never let my right hand know what my left hand does. I'm perfectly willing to mislead and tell untruths to win a war," at the least. And as clearly stated on the greatly watched and much respected 2014 PBS documentary on the Roosevelt family, it was noted - during the year 1941! - that all his life, crafty FDR had "always loved secrets that no one else knew." He was good - almost obsessively so - at collecting and keeping these hushed secrets (like the atom bomb program), and meanwhile closing off his true thoughts and plans to just about everyone, historians reinforced on the television mini-series. Thus, most Americans were genuinely stunned to learn about the true nature of Mr. Roosevelt's health and marital infidelities in the years after his death.

Keeping important, juicy information very private - utilizing outright lies at times - was standard behavior to Franklin Roosevelt, his loyal aides, and somewhat to the willing Washington press corps that followed him around the White House and around the country. If they even tried to breathe a whisper of either of these two listed intimate personal secrets of the president, for example, there would have been hell to pay, thus they were never explored and exploited (in his lifetime). So we can clearly see that in 1941 America, if you were told to keep something secret by your president or a top official in your government or military, you did as you were told. Purposely covering up explosive issues was mainstream and unchallenged. Times have obviously changed, but morals and mores were much different in the 1940s. Basically, whatever a president wanted back then, he got. And FDR was harshly labeled by his critics as the worst offender of this description, dubbed a "dictator" and a "king" for his presidential power fully exercised, sometimes felt by some to be beyond what a democratic system should allow.

Thus we can see that the unvarnished truth about MO41 never had a chance to "go viral" in the 1940s, not with great military concealment, atomic research restrictions, a muzzled press, a frightened citizenry, and a chief executive who gleefully loved keeping intimate and shocking facts to mainly himself. And it has barely leaked out ever since... until *now*.

CHAPTER SEVEN

MO41 History Summation

"My feeling is that the Missouri incident in 1941 ... did indeed occur."

When Charlette Mann kicked off the MO41 story with her letter in 1991, it landed in the grip of UFO researcher Art Fowler. He looked into the matter, and brought in cosmic crash investigator Leonard Stringfield. After reading Leonard's mention of the Cape Girardeau alien accident in his rather obscure '91 book's chapter, another noted UFO examiner delved into the topic. Famous UFO crash researcher, media figure, and nuclear physicist Dr. Stanton Friedman had dread the leaked Twining "White Hot" report mentioning "the recovery case of 1941" and finally pieced together its clues with the Huffman/Mann account in Stringfield's book. Critical of Stringfield's research efforts, feisty Dr. Friedman personally traveled to Cape Girardeau to investigate, interviewing friends and family of Reverend Huffman and his granddaughter, and his former employers, the Red Star Baptist Church (formerly Tabernacle). "Clearly this was a man of some substance," Friedman concluded of the late William G. Huffman. Building a case for authenticity, intrigued Stanton consulted with researchers Ryan Wood and his father Robert, who were of course accumulating their own impressive files of leaked government documents and still other UFO evidence.

Learning all of this, *"Earth Files"* radio/internet host Linda Moulton Howe also got hooked. She dug into the case, uncovered more angles, and brought the story to the national airwaves via broadcasting host Art Bell's *"Coast To Coast A.M."* syndicated radio program in the late 1990s and early 2000s.

As we have seen, in the late 1990s some shocking typed-up secrets were meanwhile passed along by retired U.S. espionage agent Thomas Cantwheel, sent in the mail to a private, California-based UFO investigator named Timothy Cooper. Tim was the son of an Army Air Force officer Cantwheel may well have known while both were in the service. As mentioned, Mr. Cantwheel had once worked for General George Marshall in the Army Counter Intelligence Corps' "Interplanetary Phenomenon Unit," and then in the Central Intelligence Agency for James Angleton (for fourteen years), before the director's ouster there. Cantwheel told Cooper in his summation letter his most amazing secret: *"There was an aerodyne recovered in 1941 that crashed in southwestern Missouri* and the one captured in 1942 in Louisiana." Bombshell confirmation! Despite the minor error of saying "south*west*," here we have very clear and exciting support of the MO41 affair from someone who should know! And what is more, Tim Cooper's Air Force father was once a military investigator of unexplained U.S. aerial crashes, and knew about the classified subject of UFOs and ET visitation. So Tim went to his dad and asked about the subject of a possible "unearthly visitation" accident and recovery in Missouri in '41. According to author Ryan Wood, Tim was told by Master Sergeant Harry Cooper that *aliens really did crash-land outside of Cape Girardeau, that Cantwheel's claim was quite true.*

A bit later the aforementioned Nathan Twining "White Hot Intelligence Estimate" was indeed leaked and that further cemented MO41 research of the above investigators, with the shocking document's twin references to *"the Missouri discovery of 1941"* that was *"extraterrestrial in nature."* And then around this time we have the exposed Otto Krause remarks recalled from 1962, about how atomic bomb technology – via a navigational propulsion device, supposedly - was recovered and copied from an unusual air crash *"in Missouri in 1941."* It was expertise and hardware not from this world, considered "more Top Secret than the atomic bomb itself." This claim seems backed up by the FDR Oval Office memorandums and Twining report listed herein. According to an internet investigator, Krause had stated most interestingly and initially to an intrigued government-hired photographic expert: "There was a retrieval of a disc *before* Roswell, back in the early 1940s, in the Ozarks." Well, Cape Girardeau isn't considered *quite* by most to be *"in* the Ozarks," but very close, so we know what Otto meant now. The once-hushed story of the millennium, that's what.

Mr. James Westwood, a professional detective from Virginia, learned about MO41 in the mid-1990s, mostly from Leonard Stringfield's fact-filled UFO crash book. The private eye became greatly intrigued and traveled to Cape Girardeau to dig into the digital story files at local newspaper office. He spoke to area citizens and even placed an ad in *The Southeast Missourian* asking anyone with information on the case to kindly come forward. Westwood's determined efforts were subsequently examined in a lengthy *Missourian* article by reporter Peggy O'Farrel in July of 1998. The Cape-based newswoman explained the alien saga only by Westwood's comments and by lifting direct quotes from Charlette Mann aired in Stringfield's '91 book. While not really advancing the story with fresh insights or previously unheard-of information, O'Farrel noted a possible source of further facts that remained sealed: Leonard Stringfield died in 1994 without releasing his research notes, sources, and methods. And his family wasn't allowing anyone to dig into

Leonard's bulky files containing "hundreds of reports of UFO crashes and retrievals," as O'Farrel put it. Evidently the Stringfield family's closed-door policy still stands, thwarting dedicated researchers and seekers of the truth on MO41 and other enticing cases to this day.

Although Westwood and O'Farrel did not stir surviving eyewitnesses – if there were any left by 1998 – to come forward publicly, many citizens with open minds began to thirst for more data. They began to absorb more related material, ask questions, and spread the word. Some people began to note the striking MO41 similarities to the Turner Holt revelations, as described by his two dutiful daughters in Ohio in 1999. A rounded, damaged, metallic disc. Three dead gray alien bodies of the same size and shape, kept within three glass jars. The special storage room well down under the basement of the Capitol Building. Everything was adding up nicely as the new century arrived.

Ryan Wood authored a thought-provoking investigative article in 2001 on the Missouri alien crash for Mutual UFO Network's monthly *"UFO Journal."* The editors of the serious paranormal chronicle added that their own source, Dr. Robert Sarbacher (someone "close to government research and development"), *confirmed that previous ET crashes and controlled recoveries like this by the American military have indeed occurred and that the Missouri case may well be the real deal.* Ryan interviewed Charlette Mann at great length for this printed story and he emphasizes to this day that Charlette hasn't embellish or alter her own recollections, and to her credit openly admitted when she simply did not know the facts. Her believable family account is the main attraction within a detailed chapter on MO41 within the pages of *"Majik: Eyes Only,"* the book on UFO crashes Ryan took years to accumulate, shape, write, and then publish in 2005. It is still considered a "must-read" by those interested in serious answers to the question of life beyond Earth, visiting here and finding tragic consequences.

Through more contemporary internet websites and paranormal podcasts; radio and TV interviews with Charlette Mann; and other UFO book mentions, the MO41 saga started to take off on a national scale in the new century, aided greatly by the 2003 SyFy Channel program *"The Secret (Evidence That We Are Not Alone)."* This television documentary slickly summed up the results of Robert and Ryan Wood's quiet, ongoing citizen investigation. The highlight of *"The Secret,"* arguably, was its graphic William Huffman account turned into a well-made recreation (although the uniformed army officer shown in charge of the countryside ET debris scene appeared older than most generals in retirement!). The startled Cape Girardeau Christian preacher was ably shown clutching his Bible, startled by surrounding Army personnel amidst various onlookers, dead aliens, vintage car headlights, and grassy nighttime confusion. The brief, simple scenes overall seemed spot on.

"The Secret" aired repeatedly thanks to positive ratings and viewer feedback, and was released on DVD in 2007. For the special documentary program, directed by Gary L. Beebe, the Wood father and son team imported noted and respected UFO authors like Jim Marrs, Stan Friedman, Tim Cooper, and the esteemed British UFO expert, lecturer, and author Timothy Good for his somewhat outsider's assessment of MO41. A highly educated and cultured man, Mr. Good weighed in on the Midwestern extraterrestrial crash-landing situation, stating firmly on camera: "All this business got started during World War II," which was raging in his native United Kingdom in the Spring of '41. "My feeling is that the Missouri incident in 1941, as relayed by the Reverend William Huffman to his

granddaughter, did indeed occur." While this strong
statement of support gets the Show-Me State claim slightly
wrong - Reverend Huffman never spoke *directly* to his
granddaughter Charlette Mann about the amazing
happening - it shows that someone from beyond American
borders and culture who seriously investigates such cases
with a healthy dose of cynicism *supports the Missouri
allegation with certainty*. Mr. Good summed up MO41
elsewhere as "the earliest reported (crash) event, as far as
I'm aware." One again, it's the granddaddy of all cases, the
very first of all American alien accidents, even to those that
enthusiastically research these matters.

American podcast producer Lanette Lee, from *"Beyond the
White Noise,"* hails from Sikeston, where she and others
first discussed the alien crash a full "thirty years ago," she
said, *long before* the Charlette Mann letter that kicked off
the national media interest. In the mid-1980s, some of
Lanette's classmates in high school talked about the rumors
they'd been told growing up in Scott County. {This
author's step-mother recalled she also first heard of MO41
also while in high school, this as far back as the 1960s, in
Cape Girardeau.} Lanette made sure to relay as much
information as she could with her internet broadcasting
friends and technology in 2015, taking the case quite
seriously.

Perhaps other, older residents of Sikeston recall when
Ronald Reagan came to their town to campaign in 1964, or
when stumping in Cape Girardeau in the 1970s and again
as America's commander-in-chief in 1988. And then
remember when President Reagan made some eyebrow-
raising remarks in public speeches during the safe second
term of presidency about our planet's leaders and citizens
banding together if they faced "an alien threat."
Coincidence? Perhaps, but director Steven Spielberg
revealed in 2011 that Reagan turned to him after a 1982

White House screening of "*E.T. The Extra Terrestrial*" and
stated without grinning, "There are a number of people in
this room who know that everything on that screen is
absolutely true." Spielberg was so obsessed with long-
limbed, big-headed aliens he included them in many of his
films, including a 2007 one in which a character states
authoritatively "There were other crashes before Roswell."
Ironically, Spielberg may well have flown with his wife,
actress Kate Capshaw, to Cape Girardeau in the mid-1980s
to pick up her Chaffee-based daughter (by her first
husband, the son of Chaffee's mayor). Did they know
about MO41, and even go to or right by its crash site, or its
supposed memorial nearby?

Reviewing the accumulated 1941 crash evidence with Stan
Friedman and Ryan Wood, Charlette Mann added for truth-
seekers in a Tyler, Texas, February of 2008 interview on
KLTV, "I have not forgotten holding that paper in my hand
{Nathan Twining's "White Hot Intelligence Estimate"
detailed in Chapter Two} and realizing that my family's
story was real," adding, "I don't have to convince anybody
{else}, I just have to know what's true for me." She's still
quite unshakably firm in her belief of her grandfather's tale,
and has never asked for a dime in return for telling it.
Investigative reporter Gillian Sheridan of KLTV in East
Texas recorded the interview and could not find fault with
the MO41 saga, either. Neither could Cape Girardeau's
KFVS with their two-part televised news report on the
story, also in 2008. {For a while, the popularity of these
reports led to them being re-aired on YouTube, although at
this point some seem to have been taken down for
copyright entanglements. Other YouTube videos on MO41
are still available.}

As the first decade of the new century passed, the MO41
storyline was included in a History Channel program called
"*Hangar 1*," which gave more information to viewers

thanks to the opening of the Mutual UFO Network's files on the case. Phillip Reynolds stepped forward, on what his grandfather the Cape Girardeau fireman allegedly related of "that night in '41." Additionally, more and more UFO books have been popping up in the past few years featuring data highlighting the '41 affair, which has been discussed within still more websites and chat forums online. For a while, at least, *"The Cape Girardeau UFO Crash"* became an official listing in Wikipedia, with different sources listed for readers to click on to learn more, including one by - of all media outlets - *"The International Business Times."* That's because that particular publication ran a 2013 article on *"The Top UFO Crash Cases"* of all time, by investigative writer Sigrid Salucop. MO41 was placed – along with two obscure British crash cases - *at the top of the list*, even above the more famous '47 Roswell controversy!

The eye-popping details in the *Topix* disclosures by "Just Wondering" and "AllSouls" tantalized in 2013, as did shortly thereafter the riveting information in Linda Wallace's family history with the Sikeston angle of the alien crash recovery process, within the e-book *"Covert Retrieval."* About.Com, the online site that tries to answer compelling reader questions on a wide variety of subjects, gave a thumb's up to MO41. Billy Booth, the website's "guide" for explaining Missouri UFO crash inquiries, explored the topic's facts and claims, then in his online summation remarked that he'd like to see more information to help prove it to be 100% authentic, but resolved in the conclusion: *"I personally feel that the crash occurred."*

George Noury, the famous and popular nationally syndicated radio host, has explored MO41 on his program at times, being often based out of St. Louis and having a grandchild who attended downstate Cape Girardeau's university. Dave Schrader, a Minneapolis-based radio host, has also covered respectfully the Missouri UFO crash, including a special two-part interview with Linda Wallace in May of 2015. At first a national radio host and star, and now on satellite radio, paranormal pioneer Art Bell has also covered much MO41 ground on the air. Mr. Bell and Mr. Noury often relied on the helpful investigative scoops of regular radio contributor Linda Moulton Howe, as mentioned. She interviewed Charlette Mann at length and also came to the firm conclusion that MO41 was the genuine article. In mid-2014 Linda summed up quickly but succinctly her updated feelings on the growing case, within an e-mail: "I have no doubt that Cape Girardeau crash in 1941 occurred, with a craft, neutronic propulsion system, and bodies. Two dead and one alive," at least at first.

So the riveting data and the credibility just kept on piling up, slowly revealing the many facets of the slowly-polished MO41diamond, no longer "in the rough," as they say. Perhaps the best overall informed MO41 source, Ryan S. Wood, for over two decades now has undertaken extensive research and painstaking testing in the field of (leaked) documents regarding secret U.S. spacecraft and alien body recoveries, most particularly when it comes to the compelling Missouri case. Mr. Wood bolstered the MO41 storyline by accumulating many leaked government documents, airing them in his amazing MajesticDocuments.com website, including five pages of a private report the imminent physicist Dr. Edward Teller wrote for President Reagan in the 1980s. In this interesting reprinted document on Ryan's site, Teller mentions the need for the scientific community to recognize the needs of "national defense" and act upon "actual danger to the United States." As an example, Dr. Teller stated this supportive blockbuster: "A danger was perceived in 1939, only two years before a UFO was captured, and Pearl Harbor." *Another bingo!* More once-secret insider backup that the 1941 incident was of a genuine alien origin. *and* that Ronald Reagan was evidently familiar with the unexplained remark's factual basis when the great nuclear scientist Teller dictated the typed report but didn't bother to elaborate on this statement. In other words, since the president had already been briefed on the Missouri crash recovery saga, he - like President Truman in 1947 - didn't need further documented explanation. The Teller letter's sentence also ties in the scientific world of physics with knowledge of the craft; Edward was a D.C.-based physicist in the early 1940s, likely to have been consulted on the recovery's properties during the war.

Ryan Wood had help from supportive researcher Cooper, who received many UFO documents from his inside American intelligence sources, and shared them cautiously. Early on, Ryan convinced his mature, educated, and worldly father, Dr. Robert Wood, that the MO41 evidence was valid and solid. "I believe it was the U.S. Army that retrieved the Cape Girardeau, Missouri, {UFO} crash," Bob Wood stated clearly in an extensive 2008 interview, in case anyone doubted his informed position. Along with his dad's help, Ryan worked the case from many angles, including open-mindedly importing a psychic, Annette Martin, who "confirmed the Huffman-Mann story and provided new checkable data." Ryan traveled to Cape Girardeau, checked local record files (what little remains from 1941), and even did some digging at a few sites. He also brought in a so-called "remote viewer," a rather specific type of psychic sleuth, a man named Joe McNoneagle. Ryan admits that Joe's input was "of speculative value without confirmation" and not supported by physical data - not yet anyway - but he informed readers that Joe insisted he could sense that there exists "small pieces of wreckage still at the crash site." Obviously actual digging and finding artifacts in this previously publicly-unexplored crash site would be of great historic and archeological value, overall a tremendous help in proving the MO41 case for remaining skeptics.

It seems beyond belief now that all of the aforementioned diverse investigators, researchers, and authors herein cautiously studying MO41 are either deluded or lying. Instead, *they* have concluded that something *very* substantial, momentous, and otherworldly really did happen in April of 1941. Perhaps to believe otherwise is to feel that all of these diverse, educated people are somehow honestly mistaken or just plain stupid, gullible, or brainwashed, but that certainly strains credulity. The skeptics and the cynics will always want more documented evidence and hard, tangible proof, and that is their right. We all do. But consider this: *no investigator on the MO41 case has come back with a negative, disbelieving report that has destroyed the credibility of the allegations involved.* No one has said "I've looked into this and it turned out to be a bunch of nonsense!" In fact, just the opposite is true. The more one learns, the more one understands MO41 is the believable truth, and that the April 1941 events are perhaps the most amazing and exciting nearly-untold saga in American history, *perhaps in all of human history*, given the full scope of its impact on mankind's science and technology today. Thus, the citizen investigation rightfully continues. If only the American government and military would 'fess up and publicly join in!

If you know anything not already covered in this book about MO41, and would like to get it into the public record, please write or type it up and send it to this author, in care of the publisher. Please do not include hypothetical theories, distorted opinions, lies, or disinformation. A second volume of fascinating, factual evidence for this amazing case is being pieced together and will hopefully act as an impactful sequel of sorts to this book someday.

BONUS CHAPTER

Shock Claim: FDR Killed Himself

"He put a pistol to his head and pulled the trigger."

It seems admittedly at first utterly ridiculous, but there's a fascinating yet shocking claim gaining steam - mostly through the internet after the year 2005 - that President Franklin Delano Roosevelt did not die as the media and history books record. That instead while on vacation in Warm Springs, Georgia, he carefully waited until the 12th of April, 1945, and found some private time, nearly alone. Perhaps depressed over his declining health and other pressing personal and professional problems, FDR supposedly pulled out a hidden gun and shot himself through the skull, then died of a cerebral hemorrhage resulting from this grievous wound shortly thereafter. The amazing allegation goes against every official version of the famous president's end-of-life story, and might be difficult if not impossible to prove. But it might also be true.

One substantial matter that concerns us here, within the MO41 storyline, is that all of this took place on the exact fourth anniversary of the extraterrestrial spaceship crash. The tragic Georgia events that day in '45 made the Missouri farmer-turned-politician, Harry S Truman, the new American president. If Franklin Roosevelt really did kill himself in this described way, it must be asked: did he plan it deliberately for the MO41 anniversary date? And why? After all, he could have chosen any other day, and he had been in Warm Springs since March 29th. Did MO41 somehow have an influence on his decision to take his life? Or was it all a coincidence?

By taking a slow train in late March to Georgia, FDR had successfully separated himself from Washington D.C. and familiar figures like his wife and family; his best friend; most of the press; his cabinet; his personal physician; and nearly all of his staff, minus a few exceptions. This allowed him to find sufficient time to be alone, evidently with a loaded weapon, with only few traveling White House press corps reporters and photographers, Secret Service agents and Marines hanging around the outskirts of his guarded Warm Springs compound.

An Associated Press report - found on the front page of Cape Girardeau's newspaper and thousands of other late April 12th editions around the world - told of the initial announcement of the president's death, issued by White House spokesman Steve Early. At 5:35 Eastern Time, Mr. Early dialed up all three major news agencies at one time in a quick "conference call" and declared only that "the president died suddenly this afternoon," adding simply, "I have no statement." Oddly, there was no official inkling just how and why FDR died, no details, not yet at least.

One intrepid reporter soon wrote that he learned that the president "died in his bedroom" in his cottage situated "atop Pine Mountain" in Warm Springs, but the manner of his demise was not explained. After that, news stories began rolling in that included the official claim that FDR had suddenly complained of substantial occipital head pain, fell unconscious while seated "in his chair" in his living room and was carried back into his bedroom. There he lingered on his "maple bed" for hours, comatose, worked on by frantic doctors. The White House version of events is that the weary Chief Executive finally passed away peacefully, almost nobly, in his bed, in no pain at all, the brain hemorrhage too great to recover from. This nonviolent claim has been accepted as fact by history. But was it the full truth? There was no inquest or legal investigation, nor any press examination, into the circumstances afterwards, just an embalming and a funeral. Everyone simply accepted what the president's aides told them. There may well have been an autopsy on the body as it lay on the simple bed for many hours in Warm Springs, but all records of it have been officially and mysteriously "lost." Franklin's son-in-law, Curtis Dall, issuing a tell-all book in 1970, admitted his own suspicions about the claimed circumstances surrounding the president's sad end, stating he heard several versions "that differed considerably" and that the official story sounded "canned" and "well-polished for specific political effect." Which tale was the truth? There can be only one, and even FDR's ex-son-in-law seemed confused and dubious.

The subject of an actual F.D. Roosevelt suicide was first publicly raised in an obscure, hypercritical, thirty-page manuscript, *"The Roosevelt Death: A Super Mystery."* This bold but nearly-unnoticed 1947 screed mentioned the curious circumstances of the president angling to be alone in Georgia, and then the unusual rumors of how he "put a pistol to his head and pulled the trigger." The overlooked book's author added, "there is much to support this theory." Back in the late forties, and for the next sixty years however, Americans did not want to hear such tales. It seemed outrageous, baseless, and traitorous, and perhaps just plain loopy. Today, the mood of the country is much different, more open-minded and tell-all, and the media more is vigilant and less silent. In most cases suicide attempts and depression are better understood now and treated with preventative study, counseling, therapy, and medication. Things were obviously very different all around in conservative 1945.

Franklin Roosevelt seemed to live his life in precise four-year segments. For instance, he attended high school for four years, then college for four years, then law school for *almost* four years. He won a pair of two-year state congressional terms, which would have lasted of course four years total until he left early to work in the Wilson administration. He conducted a four-year extramarital affair with Lucy Mercer. Mr. Roosevelt later served four years as New York's governor, then won four four-year terms as U.S. president, the last of which he was unable to fill out due to his sudden demise. That regular block of four years weirdly yet fittingly seemed to be intrinsic and important in Roosevelt's mind and lifetime.

Consider an insightful modern-day observer of the suicide allegation, online contributor Jay Evenson. He wrote on Amazon.com that there was a unique war angle weighing very heavily on the ill Roosevelt's mind in his final weeks, days, and hours. That this was a great contributing factor in the president's decision to take his own life. An important atomic weaponry test was run on the previous day, April 11th, and the technology responded well, the president was soon informed. Its destructive power was genuinely frightening. Mr. Roosevelt "did not want to go down in history as a "war criminal,"" Evenson declared online in 2014, and a *huge* decision was creeping closer and closer to overwhelmed FDR on whether or not to actually use it. Unleash the A-bomb and wipe out possibly *millions* of German and/or Japanese lives - many of them civilians - or risk great personal guilt and domestic criticism later of having held back and costing thousands of American lives in countless more bloody battles. *If* this was truly a great burden pressing hard every day on the polio-stricken president, MO41 once again plays a bit of a role. *The Vannevar Bush-guided atomic weapons program was augmented thanks - supposedly - to the alien technology propulsion device copied from the Cape Girardeau ET crash materials (*as privately mentioned by Otto Krause and other sources mentioned herein*).* Every day that passed, the need for firm, decisive action on this explosive matter with a MO41 angle burdened Mr. Roosevelt further.

Increasingly poor health also *had* to have greatly impacted the president's fading morale and attitude by the early spring of '45, no question about it. In addition to hypertension and very high blood pressure; "pretty severe headaches" and chronic sinus congestion; and hand tremors, the president suffered still other medical dilemmas. He experienced periods of poor or even *no* concentration, likely due to epilepsy; a dangerously enlarged and congested heart; weight and appetite loss; a possible forehead tumor or melanoma skin cancer problem; and painful infantile paralysis. In his last days alive the president complained of intensified polio pain in his joints (mostly his knees), making standing nearly impossible, and a low grade fever. He may well also have had prostate cancer surgery the previous November, as rumored, with lingering effects still present, if not spreading. All in all, with great due respect, FDR was a *mess*, lucky to have lived as long as he did. And he knew it most of all.

Biographers describe this ill, pale, and frail President Roosevelt as locked into burdensome, weary war frame of mind and feeling the daily pressures of leadership taking its emotional and physical toll on him. The stressed Commander-in-Chief compounded all of his known health problems with a habit of smoking up to four packs of Camel cigarettes per *day*. Supposedly this was allowed in order for FDR to deal with the tremendous burdens and stress of his high office, but it certainly only worsened his health woes, via shortness of breath and coughing due to congested lungs.

It should also be mentioned that the polio-plagued president who repealed prohibition also regularly mixed and sampled his own afternoon cocktails, at times more than he should have as the evening wore on, biographers note. It undoubtedly helped him dull his intractable pain. Franklin had still other dilemmas and dramas to deal with; he was familiar with cousin Teddy Roosevelt's son, Kermit Roosevelt, for instance. Kermit suffered an alcohol and depression problem. His sudden and shocking death may well have played a key role in Franklin's, perhaps even serving as an eerie template.

In 1943, FDR had assigned Kermit Roosevelt to a special post in Alaska during World War II. Less than two years prior to April of 1945, Kermit struggled mostly alone with his personal demons as the Japanese military didn't show up in his assigned region. There was actually little to do. *Separated from his family and friends in remote Alaska, Kermit went into a room alone, shut the door, and put a gun to his head, then pulled the trigger.* He died on the spot, slumped over, a bullet in his brain. This scandalous Roosevelt family event was covered up and the public - and even many relatives - were told that Kermit died of a "sudden heart attack." It took many years for the full truth to come out.

The Roosevelt family's successful '43 suicide cover-up had to have been weighing on FDR's mind and memory as he found himself nearly alone in South Carolina, resting and rehabbing in a friend's mansion in April of 1944. Only a small staff was permitted with him. His personal assistant, Daisy Suckley, kept a diary of that troubled private time for FDR, and noted in it on April 4[th,] 1944, that the president "is seriously thinking of having a *gun* brought to him." Incredibly, Franklin Roosevelt really wanted this done, as the third anniversary date of MO41 approached. Supposedly he desired a firearm because, Daisy wrote, he wanted "to shoot all the disagreeable people" he was encountering in life, which of course is nonsense. The president may have used this as a joking excuse for possessing a loaded pistol since he had to come up with *some* sort of "cover story" to mask his real reason for ordering a handgun. He knew the weapon could someday be his "out," to act as a way to permanently relieve his suffering, just as his relative, Kermit, took similar action in Alaska nine months earlier. Now a year after this remote Carolina idyll, convalescing in Warm Springs, Georgia, Franklin was again (nearly) alone in a room, apparently with his gun, away from family and friends, his physical distress and emotional frustrations mounting since his arrival there on March 29th.

Another similar scandal that touched very close to home was the death of FDR's beloved secretary and friend, Missy LeHand. She suffered a stroke in June 1941, and fell into such ill health and depression that she attempted suicide at Christmastime, then a second time weeks later while staying in the White House (by setting herself on fire!). By the time sickly Franklin was pondering his fate in Warm Springs, distraught Missy was dead. So was his beloved mother, Sara. So was his closest advisor, Louis Howe. So were a number of other old friends and family members, including Kermit, of course, and also depressed alcoholic Hall Roosevelt, FDR's brother-in-law, and *his* son. And Teddy Roosevelt, Jr., who died in Europe's war theater in 1944 (after having been activated for duty in April 1941, of all possible times). And then there was Franklin's father-in-law, the First Lady's papa; he died of injuries resulting from his own tragic suicidal attempt in 1894. Add to that the gut-wrenching war and ongoing genocide that was killing *many millions*, around the world. Sudden, violent deaths and somber mourning periods were not uncommon for the president, and for the nation, in the 1940s. They were the norm. Even the odd extraterrestrials who were found in Missouri in '41 died sudden, violent deaths.

One final key to understanding FDR's Thursday, April 12th, 1945, suicide motives was his very real superstitious triskaidekaphobia, where he refused to do much at all on 13th of any month. By mid-afternoon his chances of getting alone "me time" before the 13th arrived were dwindling. He *had* to act soon, since he apparently very much feared Friday the 13th!

The popular president was not due back in Washington until April 20th, so there was a full week of the Warm Springs idyll to come. He had tentatively planned to travel by train later in the spring to San Francisco, to attend a new United Nations conference, but suspiciously had not bothered to complete that rather loose but important itinerary. His only idea was to arrive by train there, give a speech, and leave town immediately, he told an aide.

Around one o'clock on the afternoon of April 12th, 1945, at the Little White House main cottage, the president signed some documents as he made polite conversation with a Russian-born artist who was began painting his portrait in the living space of his personal cottage. His mistress Lucy Mercer sat nearby. He wore a gray suit with a crimson neck tie, appearing wane and weary, pale and ponderous. Some aides were in and out that day, some preparing for a simple barbeque picnic on the grounds around 4:30. From here, notable variations of the story abound.

According to the official version of the tragic afternoon, Mercer's friend, Elizabeth Shoumatoff, was the artist working on her now-famous watercolor design while the president's secretaries Grace Tully and Laura Delano were outside the house, elsewhere on the estate. As possibly was the president's nurse; one version said she was evidently waiting outside the cottage for the signal to come back in, another claim was that she was actually seated near FDR all afternoon. Mercer was settled on a nearby sofa, chatting with personal aide Daisy Suckley, who crocheted. FDR's valet and his assistant were supposedly setting out lunch in the nearby dining room. Shoumatoff said she was told during the painting session by Roosevelt himself that he would sit for only fifteen more minutes. Supposedly Franklin soon afterwards stated "I have a terrific headache" and closed his eyes. He allegedly dropped his papers to the floor, fainted, and slumped forward in a chair before a table he used as a desk. Another account says he slumped *backwards*, while another claimed it was FDR himself who fell unconscious to the floor himself, not his papers. Still other accounts - as we shall see - say FDR had excused what little company he had that afternoon and did not collapse in his chair by any *natural* causes. As the accepted official claim goes, after FDR passed out, his valet and the assistant were swiftly summoned and they carried the president from the padded chair and into his nearby bedroom.

FDR's vacation physician, Dr. Howard Bruenn - a respected cardiologist - stated he was urgently called and arrived minutes later. {Admiral Ross McIntyre, the president's usual physician, was back in D.C.} The doctor gave a 1994 interview and remembered he rushed inside to find FDR slumped over, still seated at his living room table/desk. It was *Bruenn* who helped the president's valet carry the leader back to his bedroom bed, the doctor recalled. He then examined unresponsive FDR and found him near death, laying silently with shallow breathing. The desperate examination somehow revealed a massive stroke, or cerebral hemorrhage, within his brain - but what caused it? As mentioned, the great man never recovered consciousness. His heartbeat and breathing eventually stopped during futile resuscitation attempts, his vital signs monitored closely. Despite an injection by Dr. Bruenn to restart the heart, it was too late. Franklin Delano Roosevelt was quietly declared dead at 3:35 Eastern Time. Distraught presidential aides, secretaries, Secret Service guards, the nurse, and trusted servants rushed in and out of the small house during this time, unable to do much more than worry and cry, console each other and attempt to reach the First Lady and the rest of the "next of kin" Roosevelt family, plus some White House staffers in Washington. Mercer and Shoumatoff packed quickly when told Eleanor was on her way from D.C. in an airplane; the pair left hastily by automobile. One historical author claims it was while traveling that this duo first learned over the radio of the president's death, which seems dubious at best. Why did the two women flee so quickly if FDR had simply "fainted" as one story claimed, or had a stroke but was still alive? What kind of supportive friends were they to just take off when their beloved president was still alive and needing comfort?

Now consider another account, re-printed within *"The Barnes Review,"* an internet website dedicated to selling books covering lesser-known accounts of historical subjects. The 2006 reviewer stated his personal research showed that some sources claimed President Roosevelt specifically "asked his nurse to leave him alone with the artist and not to return until he summoned her. She complied," and dutifully stepped outside. "After some time she saw the artist {Shoumatoff} leave the building and was waiting for FDR to summon her. A while later, she heard the sound of a pistol shot." Naturally the nurse rushed into the building the president had ordered vacated. She found Franklin "slumped in his chair," with "clearly a bullet hole through his head." The attending nurse told others that Roosevelt "had been despondent for some time and often spoke of making grave mistakes," which obviously weighed very heavily on his conscience. {Where Lucy

Mercer and Daisy Suckley were during all of this is unknown, and if the valet and his assistant were busy in the cottage dining room, it was unstated.}

Next, take into account a 2006 posting on a Yahoo.com internet chat board. "My English teacher told our prep school class in 1947 that he knew Madame Shoumatoff," the painter. The teacher informed all that Shoumatoff had privately confided to him that "the president shot himself and that she had seen his body slumped over his desk with a pistol" still gripped in one hand. The teacher, the online poster stated clearly, "was a well-respected and honorable man and wouldn't have made this up."

In 2009 another online poster stated a good friend of his had an aunt who was the nurse in question that day, assigned to care for FDR at the Warm Springs cottage. She had been asserting to family members and the friend that "she wanted the truth to be known that FDR had shot himself with a handgun."

Such reports would account for why Mercer and Shoumatoff would panic and quickly flee the gruesome and scandalous scene that reporters might sniff out while a gentler cover story was hastily prepared. The duo would likely not take off suddenly if FDR had only "passed out" or fell ill. Yet once again, these contemporary statements aren't actual proof that FDR pulled the trigger and that it was all carefully covered up by aides, but they start to sound clearly and eerily familiar, and seemingly very believable, when taken into account with many other online forum statements that came flooding in following these opening remarks, some of which we'll examine herein.

As an example, in 2006, writer John McCarthy passed along a little-known story from '45 that mentioned "an elderly black man in a rocking chair" outside the presidential vacation home, crying his eyes out that fateful afternoon. "In his grief he stated "The Master is gone, The Master done shot hisself."" Supposedly this blind African-American was the president's favorite local musician who had entertained FDR's guests at times, and had heard the fatal gunshot coming from within the cottage. According to McCarthy, "the FBI and Secret Service grabbed the old man and removed him" from the scene, and he was supposedly not seen around those parts again.

On an Amazon.com 2010 internet bulletin board, Edmond Gunny chimed in with a helpful memory of what first drew him into this startling historical mystery. "The father of a close friend told me in the early '60s that an insurance client of his was a Secret Service agent" who was assigned in 1945 to protect President Roosevelt. The unnamed agent informed the close friend's father that the terrible truth was that "FDR shot himself in the head." Gunny claimed that he had not learned of any corroboration of this until he searched on the web and "bumped into this post," one that listed other similar compelling hearsay reports. Thus he decided to add his own two cents worth as well.

A woman named Pamela Foote responded in 2011 to Gunny's web forum remarks when she confirmed apparent Secret Service insider knowledge. "My grandmother was engaged to a Secret Service man at the time of FDR's death," Foote wrote online. The grandparent told Pamela that her then-fiancee quietly relayed to her back in '45 that truth be told, "FDR killed himself."

On Yahoo.com, a 2006 forum poster wrote that his grandfather "was a Secret Service agent assigned to the president" in the 1940s. "He was outside the door" at the peaceful Little White House when suddenly he "heard a gunshot, rushed in, and found FDR at his desk." Obviously FDR would have been slumped over, presumably in his living room chair by the simple flat table, but the grandparent reported he could still clearly see the inert

president "with a gun in his hand, and bullet to his head."
The online contributor added that his grandfather was a
well-decorated war veteran, trusted Treasury Department
agent, and experienced ex-Secret Service guard who long
ago retired but only spilled the full story of Roosevelt's
suicide when he reached the ripe old age of 95 (he is said to
have later died in 1996). The bodyguard was supposedly
"forced to leave Service" after the suicide and sworn to
secrecy over the entire shocking situation in Warm Springs.
"He was a very honorable man and wasn't one to make up
stories," the board poster concluded.

In 2008, an online forum contributor stated that his friend's
father-in-law had once been another of Franklin
Roosevelt's Secret Service bodyguards. The retiree's
troubled confession to his family was that sickly FDR had
specifically asked to be left alone at one point that day in
Warm Springs, and the agent failed to realize "he concealed
a gun under his blanket on his lap." It was then, when left
by himself, that the president pulled it out and used it, to
everyone's shock.

In 2009, an online poster claimed "my two aunts from
Georgia were dating FBI agents around the time." One of
the two aunts "told me many years later" that the pair of G-
men confessed back in 1945 that the stunning hidden truth
was that President Roosevelt had actually shot himself with
a pistol. The aunt "who still lives in Georgia" had recently
"confirmed this account." This means Director J. Edgar
Hoover knew and kept the story squelched, even though he
was definitely not a fan of either Franklin or Eleanor, yet a
fellow Freemason.

Adding to these startling but confirming statements in the
Amazon online forum in 2014, Neil R. Nelson backs up the
dreadful suicide scenario. "My father was at the White
House when FDR shot himself with a .45 {in Georgia}.
I've been waiting for years to share my info."

If the story is true, then perhaps a reporter or two or were at the White House, or who had traveled with the president, overheard the news (if not the shot itself). A press report may well have been issued from the start, revealing the actual cause of death. As Jay Evenson asserted flatly: "The first reports were radio broadcasts that stated the president shot himself in the head. The official version came quickly thereafter - that he died of a brain hemorrhage." Evenson feels the nobler version of events in reality didn't stray *that* far from the truth as "it *was* a brain hemorrhage, brought on *by a bullet to the brain*." Technically, the doctors who issued the official statement on FDR's cause of death were not lying, there *had* been cerebral hemorrhaging within the president's cranium. They simply left out the *full* truth, that the gunshot lodged in his brain actually caused this fatal process. "People benevolently minimized the cause" of death, online contributor Edmond Gunny summarized, due to the somber circumstances of the national mood. After all, Americans were fighting for their lives in battle all over the world. How would it look if its famous leader abruptly gave up and killed himself? As another poster put it: "(The truth) would have devastated the country's morale, given much comfort to our enemies," and reflected a kind of "weakness in our national image." Truer words were never spoken. It had to be covered up back then, for the good of all, and was.

It seems plausible that a select few in the media or at least the military may well have actually sniffed out the truth and initially reported it before the Roosevelt administration could muzzle them on April 12th. One online contributor stated: "My grandfather told me in the 1980s that while he was aboard a radar vessel, they received the radioed news that FDR died of a "self-inflicted gunshot wound to the head."" Not long after that stunning initial radio message, the official story was - unsurprisingly - changed. "It was revised to "cerebral hemorrhage."" Such a sad but urgent message may well have been gleaned from a radio broadcast, but more likely from a military source, possibly from within the White House, sending out an emergency bulletin over armed forces radio channels that afternoon.

Another forum source claims he heard the same thing. "My grandfather was docked in the Philippines," the online poster remarked, and "he remembers the same announcement going over the intercom." Suicide by a single shot. The author of the posting went on to describe his own mother's experience. She had long ago claimed to her family that she was "shopping in Bloomingdale's in NYC at the time of the news of FDR's death was announced." The store's shocking declaration was aired perhaps via a public address system and "clearly stated that he'd shot himself in the head. Not long after that . . . the news media suddenly changed to his death having been of natural causes." For the rest of her life, the woman in question supposedly "never wavered from this account, and had no reason whatsoever to lie about what she heard that day," her daughter added online.

That Franklin Roosevelt's body was never put on display raises deep suspicions in citizens still to this day. One example: a Yahoo.com forum poster reinforced: "My father told me something was wrong" right from the start, owing to the great man's remains not being available for viewing even by the Roosevelt children. It is apparent the First Lady demanded the coffin lid be kept shut at all times, including at the private Hyde Park funeral service, the body off-limits to even the next of kin. Why? If it was a simple cerebral hemorrhage, his subtle injuries would be contained within the cranium and remain unseen to the naked eye. Then Franklin should have been quite presentable as he lay in his casket. If he was sporting an ugly bullet wound to the head, then concealing this exterior injury for an open coffin would obviously have been very difficult, if not impossible. Also, as a Freemason, FDR should have received a masonic funeral ceremony with many members in attendance, another forum poster pointed out accurately; that he did not is very suspicious in that *masons always deny funeral services to those members who commit suicide.*

Experienced morticians were called to attend to the president's body in Warm Springs, that is a fact. They arrived there late on the 12th and their mission was mostly to embalm his remains carefully and tidy up the president's appearance, presumably for a public funeral after laying state. A team was assembled from a Spring Hill, Georgia, mortuary, located just north of Atlanta; they gathered their equipment and drove down to Warm Springs as fast as possible. One of them even handwrote a fifteen-page account of this extraordinary experience, available for viewing online today. The embalmer mentioned his team arriving at 10:40 p.m. Thursday, but not being able to access the famous form until 12:15 a.m. Friday. It seems the Roosevelt inner circle wanted Eleanor to arrive first and take charge of the situation. Everything was on hold as per her wishes. Fair enough, but this caused a huge delay in the process; from 3:35 p.m. to 12:15 a.m. there was nothing evidently done with or about the president's corpse. All present had to sit and wait for the First Lady's party, flying in from Washington after Truman took the official oath of office, with FDR's form still laying on his bed, being viewed from time to time by authorized persons. Apparently no autopsy or pathological procedure was performed at this time; at least, not a soul seems to recall this having been undertaken in Georgia.

When the Spring Hill mortuary squad was finally allowed to begin their overnight work in the guarded presidential bedroom, the undertaker wrote, the president's body was still on his bed but "a sheet was drawn over his face." That may have seemed odd, but what they found next was even stranger still. "Tied underneath his chin and over his head {was} a strip of half-inch gauze." This conveniently covered a portion of the skull. Supposedly this gauze-tying was done to "close the mouth." This happened mostly due to rigor mortis anyway, making the strip pretty unnecessary. Obviously this gauze would also have been utilized to cover up the cranial bullet wound and any possible signs of emergency surgery undertaken there. Such an injury was never mentioned in the report, likely on orders. However, the same pen-in-hand mortician tellingly scrawled: "We, of course, had to use heavy cosmetics to cover the disfigured areas." What part of FDR's body was so badly "disfigured" that it required "heavy cosmetics"? We can sure guess now. {The same Spring Hill employee did not mention in his notes any sort of pre-embalming autopsy cutting or "Y" incision nor any stitching up of the famous corpse.}

Also, it was noted in the embalmer's report that a Marine guard was present at all times as the team tried to do their work on the body, but the mortuary workers were not even allowed to comb FDR's hair. That chore was strangely left up to the president's valet, who was present just after he died and helped carry the great man to his bed, so he already knew the full extent of his cause of death.

When the embalming, tidying undertakers were finally done at 5:45 a.m., the report claimed, they finally called in Mrs. Roosevelt, who had not seen the body during the mortuary process, but just before. Eleanor was allegedly quite pleased at what she viewed of her spruced-up late husband, who now lay in a Spring Hill coffin. If this is true, why then the ensuing decision to keep the casket lid closed at all times? The Spring Hill mortician also noted that the few to view the body in the casket next after

Eleanor around sunrise was "one of the Marine Guards and Secret Service men."

A suspicious aspect of the first-hand undertaker's account is highlighted next: "After we completed our work, the Government took charge." Really? What was anyone in "Government" (a.k.a. the Secret Service) doing with FDR's prepared dead body in private, from 5:45 to 9:15 a.m.? Something quite possibly suspicious, conspiratorial, and very controversial? "At 9:15 the casket was placed on the hearse" and driven off the grounds under heavy escort, to the train station, the mortician recalled in his documentation. But why the three and a half hour gap? The First Lady and her party were all present and ready, if not eager, to go home with a full casket as soon as dawn broke.

The Spring Hill mortuary notations fit in well with an online web forum post from 2008. A person there wrote that his "best friend worked for a short time at a mortuary in Atlanta that processed" the Roosevelt cadaver. The friend was informed "by some of the older black workers that the body had a single gunshot wound to the head." If such a story is true, then the Spring Hill group obviously did discover the disturbing head wound, gauze strip or not, and privately talked about it with co-workers upon their return.

On Amazon.com's forum, a 2009 poster named "Leonidas" claimed his good friend living in St. Louis knew an elderly doctor who had once told him that way back in 1945 "he assisted in FDR's autopsy." It was there that the then-youthful doctor learned to his surprise that Roosevelt had in fact "died of a self-inflicted single gunshot to the head." This was achieved when the president had pulled out and used "a .45 automatic pistol (Colt military model)." This specific weapon was concealed "in his lap robe" and that FDR carefully waited until he was alone on that fateful afternoon to end his life without a struggle or interference by others. Certainly the level of detail here in the doctor's admittance sounds like it came from insiders who were privy to the facts.

Responding to Leonidas in 2009 was a man named Curtis Carter. He stated on the internet bulletin board that the autopsy story passed along "would coincide with my doctor's account" of his 1982 conversation with "the Attending Pathologist" in the hushed FDR case. Supposedly the first thing the pathologist informed Mr. Carter's doctor was that President Roosevelt did not die of just a cerebral hemorrhage, but that the true cause was the self-inflicted cranial gunshot. "I was in attendance when the autopsy was performed on him and I removed the bullet from his brain," the aging pathologist allegedly admitted. But when and where was this alleged autopsy done?

Once at the White House funeral service, Eleanor cleared the East Room of even military guards to open the casket and see her husband one last time in complete privacy, an usher wrote later. Then FDR's firmly-shut casket was viewed only by a limited number of friends and family (and guards) with Old Glory atop it in that special room. It was surrounded by flowers and stunned mourners - and guards. The coffin was then hauled back to the railroad station and taken by train to Hyde Park as mentioned. At some point - if there was genuine subterfuge going on - the body had to have been brought to at least the grave-side coffin (where the various guards were at last relieved) and placed inside it in time for its burial near the late president's home near the Hudson River. This is presumably where it rests to this day.

One person who was not present in Warm Springs found out the whole truth and never spoke a word about it to anyone, evidently. The facts were wormed out of those present by none other than First Lady Eleanor, who by all accounts quizzed everyone present as to the true circumstances of the tragedy just after she arrived. She also at one point viewed Franklin's body alone for five minutes, before mortuary work, then again afterwards. Yet Mrs. Roosevelt evidently never gave anyone details surrounding her husband's demise, although perhaps did give us a strong hint of what really took place when years later she wrote of her feelings on the train headed north with the casket on the day after his death. {Ironically, the superstitious president who hated travel on the 13th was now being transported across the country on, of all days, Friday the 13th!} "The only recollection I clearly have is thinking about *"The Lonesome Train,"* the musical poem about Lincoln's death. I had always liked it so well - and now this was so much like it." A fascinating clue! A war-weary, seated President Abraham Lincoln was done in "with a pistol bullet through the head," as the poem goes. Thanks to this slug in the brain, unable to be removed surgically, Lincoln went into a coma, in mid-April, almost exactly 80 years earlier than Roosevelt. Abe was carried by a doctor and some men to a small non-White House bed where they monitored his labored breathing, probed his head wound, and took his vital signs, just like FDR. After some hours of anxious waiting, with aides and friends visiting, unresponsive Mr. Lincoln faded and died of his bullet wound. A train eventually lugged Abe's body home, across the country, with Americans of all backgrounds coming out to observe the sad, slow procession. Grieving citizens mourned the great man's passing, understanding that a pistol's single firing to the skull was responsible… just like with Mr. Roosevelt, evidently.

Now to wrap up on one amazing final fact. When FDR determined to end his life, and did so, where was the Missouri farmer-turned-politician whom Franklin selected as his second-in-command, to take over upon his death? On the afternoon of April 12th, 1945, Harry Truman was found down inside the U.S. Capitol Building, off limits to the public, according to historians. At one point while comatose Roosevelt took his last breaths, Truman had just entered a special downstairs Capitol "senate lounge;" he stated later he was merely waiting there for a fellow Freemason, Speaker Rayburn, in this private room. A White House phone call was responded to, summoning him. Truman then ran through the echoing old masonic Capitol crypt and underground corridors to find his hat in his upstairs vice president's office. He hurried out of the building and by car over to the White House, where he was sworn in. So of all the places Truman could have been on what one might call "MO41 Day," he was either on the very same subterranean Capitol floor the astounding recoveries were apparently once stored, or very close by! Coincidence?

Privately, to a chosen few, FDR had mentioned in his final six months alive his plan to resign the presidency just as soon as the war was over. Many insiders fretted all along he wouldn't even last that long, and this was undoubtedly a thought in his own mind at times. So of all American political figures possible, FDR wanted Missourian Harry Truman ready as the Veep, to take over when either event took place. Was this due to HST's insider knowledge of the facts on alien visitation? Soon by his own hand on "MO41 Day," Franklin Roosevelt was dead. Harry Truman was automatically the president. The next hours were a blur as Truman was swiftly and officially sworn in, then assumed staggering presidential duties and war effort briefings – including on the atomic bomb (and maybe MO41?) – into the evening hours in FDR's carefully arranged Oval Office. The day after Harry learned all of this, Truman described his shock to reporters: "Last night I felt like the moon, the stars, and all the planets had fallen on me." A very interesting and revealing choice of words considering the celestial objects that fell from the sky one Missouri night precisely four years earlier!

Truth, as they say, is stranger than fiction. And a lot more amazing and mysterious!

MO41, The Bombshell Before Roswell:

The Case for a Missouri 1941 UFO Crash

Chapter Source Notes & Trivia

CHAPTER ONE: I sent Charlette Mann a series of questions on paper in January of 2015, regarding small details, and she replied to many of them, through Ryan S. Wood, which helped me to polish this book. Otherwise there's not much else to be learned from her original, unchanging story, which is now on many online sites but the best feature more in-depth interviews with Ryan S. Wood and Linda Moulton Howe. I have found only a few small discrepancies in Mrs. Mann's account. I believe well-meaning and candid Charlette is making nothing up, but that she is simply honestly mistaken regarding a few small things. In one retelling to LMH she claimed the Huffman family was "sitting around the house between eight and nine o'clock" when the cop's call came in, whereas other accounts she has used the time-frame of "nine to nine-thirty." I split the difference herein and use 8:30 to 9:00 p.m. as a reasonable compromise. And William Huffman was *not* a preaching pastor at Red Star "church" – it technically was a "Tabernacle" - he was merely a money-raising associate pastor at that point. The facility was built in the 1920s, but in 1949, a terrible tornado ripped through Cape Girardeau and badly damaged the wood structure. It was pulled down and slowly rebuilt on the site, this time with brick and concrete. The resulting orange-colored brick church is what is at the location to this day, but it is not at all what Reverend Huffman knew and worked at. The current cozy parsonage next door was not available for the Huffmans, either; it was utilized by visiting pastors and their families when it was completed in the early 1950s. . . . Red Star's top pastor who ran the tabernacle in April of 1941 was as mentioned Harrison C. Croslin; he left the facility later in the year, however, to return to Illinois for a new position. Today's Red Star records show their congregation was shocked to learn that Reverend Croslin died suddenly one year after MO41, at home, on 4/16/42. . . . In 1941, Cape was an uncomplicated, quiet college city, with some its activities often centered around Southeast Missouri State Teacher's College classes, sports, and social events. Today it's known as Southeast Missouri State University, featuring a much larger campus, student population, and curriculum. By mid-April of '41 there wasn't much afoot. Within a few weeks, the semester would be over, and hundreds of students would be leaving town. Neighboring small farm villages had of course even less going on, the main business being farming and society along with Christian church activities. . . . W.G. Huffman was born in Lebanon, MO, having attended a Christian college in Bolivar, MO (on the southwest side of the state) in the time of the First World War. William then went to a seminary, training for his pastoral duties, apparently always with a reputation of quiet dignity and honesty. Red Star records show he successfully "Baptized seventy-eight persons" during his tenure in Cape Girardeau. His and Floy's combined, classy tombstone can be found in the *mason-friendly* Lebanon Cemetery, pictured within findagrave.com. It also covers sons William, Jr., and Wayne (who is buried in England). . . . Floy Huffman was still quite lucid but ill with cancer in 1983 and '84, eventually passing away in late November that year. It had to have been a particularly depressing time for her, care-giver Charlette, and their family; Floy's brother James died a month earlier. Floy's other brother,

Charlette's Uncle Ed, lived until 1997. They were likely not told about MO41 by the conservative Huffman brood, having no "need to know." . . . Information on Milton Cobb could be found in the 1940 U.S. Census (provided by Linda Wallace) and the Cape City Directory, and job description found online, such as in Wikipedia. Milton was listed as age 46 in the 1940 form, his wife Wilkie as age 42. . . . Two weeks after MO41, Cape firemen (again, no females allowed) would get updated new uniforms and perhaps equipment too, according to the conservative newspaper, just as a coincidence. . . . A November 2001 article on the MO41 case written by Ryan Wood for a *Mutual UFO Network* monthly publication may be the first public magazine airing the MO41 case, entitled *"Did a UFO Crash in Missouri in 1941?"* The resulting feature can still be found online, however Ryan (or his initial researcher) made some early case mistakes. He referred to the crash site as "north of Cape" which appears to be inaccurate. He stated that the city's police chief in '41 was "Marshall F. Morton," when it was actually Edward W. Barenkamp. He also named "Harold J. Shelden" as head of the Sikeston MIA, when it was in fact Charles B. Root. Ryan also named "John Cracraft" as a Cape policeman who "likely handled the bodies," when in fact I discovered a newspaper article from the month of the crash saying John Cracraft was a young U.S. Army soldier assigned to guarding prisoners at Fort Leavenworth, in Kansas, in April 1941. . . . Chief Barenkamp was killed in battle on 5/27/45 as a Seabee in WWII. Ed was an honorable Navy Boatswain's Mate Petty Officer, First Class, when he passed away. . . . Information on the falsehood created by the American media that many ("millions") in the country panicked over the 1938 radio play by Orson Welles and his acting troupe comes from Wikipedia and an October 2013 Slate.com article, entitled, *"The Myth of The War of the Worlds Panic."* . . . Information on Cape's first FBI agent - Arlin E. Jones - was provided by helpful Linda Wilkins, Public

Affairs Specialist at FBI headquarters in Washington D.C.,
via e-mail, and by a check of the 1942 Cape Girardeau City
Directory. . . . That Reverend and Mrs. Huffman eventually
bought their two-story rental house (as records show in
May of '43) up the street from Red Star, which was located
at 1301 N. Main, months after MO41 indicates his solid
standing at the venue and in the general community.
William was confident in his Cape Girardeau future and
was obviously unafraid for his personal safety and
professional career by staying in town, despite knowing
what he knew. I have been inside the Huffman house and
can testify to its cozy charm and sturdy construction,
although some residences down the street, near the church,
have gotten run down, some so badly they've literally been
torn down. . . . The 1942 Cape Girardeau City Directory
(phone book) was helpful in tracking down some addresses
and facts about life in Cape, but a current resident there
informed me it was not always entirely accurate. In the old
days, a kind of city census taker would go door to door to
gain your name, address, and phone number for the
directory, and if a citizen wasn't home, or didn't call for the
information, you simply didn't make the next year's
directory. Or the local newspaper featured an ad asking
residents to mail in their personal information if they
wanted it added to or changed from the previous year's
directory, and if you failed to respond, too bad. . . . *The
Southeast Missourian* on Saturday, 4/12/41, provided
information on the local weather conditions and forecast,
plus the Cape Girardeau fire and police personnel and
vehicular state of affairs, via Mayor W.H. Statler's remarks
in the article entitled "Mayor Sees Objectives Ahead If City
Is to Progress." . . . Some details on previous explorations
into MO41 come from fascinating web sites like
mysteriousuniverse.org, abovetopsecret.com,
theufochronicles.com, from-the-shadows.com,
ufocasebook.com, and ufoevidence.com, and more. . . . The
1941 downtown Cape police/fire station on Independence

Street is now the Cape Girardeau River Heritage Museum and can be toured Thursday, Friday, and Saturday afternoons. I was granted a private tour on 4/1/2014. My grandfather was in and out of the once-bustling building often in the 1930s and '40s, on business as a private practice lawyer and then at times Cape Girardeau's City Attorney, or just to see old friends. . . . Many facts setting the stage for the crash were taken from digitized online past editions from March and April of 1941 of *The Southeast Missourian* newspaper. For instance stories telling of the many well-attended church services and warm weather that Easter weekend. And the PAC/MIA statistics, or the engagements of Sheriff Schade and Captain Root. And the closing of the city's Army recruiting station. The fact that the cops in town worked twelve-hour shifts is from a 4/10/41 *Missourian* article entitled "Two-week Vacations For Cape Policemen Ordered by Mayor." The city firemen's late arrival at a collapsing building in downtown Cape late Saturday night was described in the newspaper's 4/14/41 issue, entitled "Two Autos Burned As Fire Destroys Private Garage." . . . . Information on Cape and Southeast Missouri State University is partly first-hand; it is the author's hometown and his college alma mater, and the same for my father, who was raised in Cape and seven years young in the spring of '41). My mother and two brothers, aunt and uncle also attended some semesters at SEMO. Helpful also is Wikipedia, plus some books found on the city's history in the local public library, SEMO University's Kent Library, and at the downtown Cape Girardeau visitor's center. . . . Information on the town's first FBI office was researched by Ryan Wood (via a March '41 newspaper article), as well as on Cape County Sheriff Ruben Schade, including Ryan's brief interview with his elderly brother Clarence Clinton Schade (who died some years ago at the age of 92). After entering the U.S. Army (Signal Corps) and serving in Europe for two and a half years following Pearl Harbor in December '41, Clarence

"Cutter" Schade came back to the Cape area and ran two menswear stores for fifty years, one in Cape, one in Jackson. He was a member of the local VFW, numerous boards, and was active with the Cape Girardeau Public Library. While doing research there I once took advantage - most fittingly - of their special "Clarence C. Schade Room," where books on local history are stored. In mid-2014 I contacted Dave Schade, the son of Clarence, and he replied that he was not familiar with the story. This makes sense in that Clarence was never at the crash site and didn't believe the wild-sounding tale when he was informed of it later, so why bother to mention it to your son? . . . Ryan Wood airs plenty on the MO41 incident and other documented UFO crash cases with his highly recommended website MajesticDocuments.com and his book *"MAJIC: Eyes Only."* . . . . Contemporary Scott County Sheriff Rick Walters informed me through Facebook that local records confirm that John Hobbs was indeed the sheriff of that county from 1939 to 1944. . . . Sikeston facts came from Wikipedia and from research by Linda Wallace as this was her original hometown, plus I've passed through town a few times. Much original Sikeston MIA (where Linda's father worked) and Parks Air College research background was also largely done by Linda, although a little more can be found on the internet, mostly through reprinted newspaper articles. She initially found two instances of cut-out and redacted articles from May of 1941 issues within her Sikeston library's microfilmed copies of *The Sikeston Herald* newspaper. This discrepancy seemed highly suspicious at first and was ably shown on her web site: seekingmoinfo.com. Later, other copies of the same editions were found and it was discovered that these missing articles were simply on mundane subjects that had absolutely nothing to do with any alien encounter. . . . Cape Girardeau citizen, enthusiastic pilot, and historical aviation buff Terry Irwin - unknown to Linda Wallace - also provided pertinent facts and opinions that helped shape parts of this chapter as well. Terry was a contributor to the *Topix* discussions; was once a member of the Cape Red Star Church; and maintains a healthy, cautious scrutiny of the MO41 allegation while researching a book on the history of regional aviation. He knew Garland Fronabarger as well. . . . Important photographic exposures would have to be slowly printed up in metal plates back in the early 1930s at *The Southeast Missourian*, and sent back to the Cape Girardeau newspaper office on a bus, to ensure the safety of the fragile process, as opposed to mailing the materials back. {"Color film and color separations" sometimes required fancier, more delicate equipment that needed an outsider's professional touch also, according to a news colleague of Fronabarger's, but since the Huffman family photo was clearly not in color, that doesn't add into our equation much here.} After 1935, Fronabarger took the lead at the newspaper to develop and print his own photos "in the printer shop" downstairs of their Broadway headquarters, with new engraving equipment having been installed. The need to send off photographic materials to St. Louis were pretty much finished by 1941. In an audio interview recorded in 1983, Fronabarger revealed he was the first local photographer to have his exposures printed up at *The Southeast Missourian*, developed from "an old, folding Kodak," referred to as a "4-A Special." {This was a virtual antique even when Frony used it, having been discontinued by Kodak in 1915!} G.D. said he acquired it for nearly nothing via his pal and mentor, local studio photographer Chester Kassel. The Kassel Studio was maintained for many years on busy Main Street, about a mile or so from the Huffman church-rental home on North Main. "I was always down there, bothering him," Frony recalled on tape with a chuckle, bonding with Chester's young friend, Jim Haman, who later bought out the enterprise. It was here where Fronabarger learned much of his shutterbug trade, with Chester showing Garland how to use cameras and develop film in his professional photography laboratory. Chester's pictures were so good, he submitted them to the newspaper at times and they were printed, right alongside Garland's as the decade rolled by. This brings to mind the unproven notion that possibly Chester was the second photographer at the crash scene, perhaps by being called to the site by Garland, or vice versa. It's also conceivable the other cameraman was Jim Haman, and it should be noted that a Cape Girardeau policeman that April 1941 was named Roy Haman, a possible relative and news connection to import journalists to the emergency wreck scene. . . . G.D. Fronabarger background info was initially produced by Ryan Wood, along with helpful newspaper librarian Sharon Sanders and other writers from past articles on Frony from *The Southeast Missourian*, where he worked for many decades. The 1983 interview between reporter Sally Owen and Frony was featured in a February 2015 online www.seMissourian.com blog, laced with many slide-show pictures the late newsman snapped and developed from around 1929 to 1986. Garland Fronabarger knew or rubbed shoulders with so many people around Cape, including my father, grandfather, and mother. Garland "grew up with Carl Fronabarger," my grandfather's brother-in-law, according to some distant relatives of mine in an e-mail. At times my grandparents hosted Garland and Carl at their house on Good Hope in '41, and later Bessie Street in Cape, where my father and his sister were growing up. Next door lived an editor for the daily newspaper, giving extra incentive for reporter/photographer Garland to show up and palaver a while. Garland and my grandpa used to smoke many a cigar. My step-mother who knew Frony said that he was the type to talk about various subjects whether anyone liked it or not, yet an intelligent man, but the smart money was on keeping one's trap shut on claims of any area crashed foreign entities. . . . Whomever the photographer was from "that night," he might have used one of several small pocket-type cameras in '41, including the Kodak "Baby Brownie Special," or the "Bulls-Eye," or perhaps the "Vigilant Junior," or even the "Range-Finder." Any of them were available to the public and produced some 4x5 photos. . . . The former sheriff of Cape Girardeau County, John Fred Hartle, was named in a 4/17/41 article by *The Southeast Missourian* as "superintendent in work being done at the MIA," which was mostly "dirt removal" to make way for construction at the Sikeston airport pilot-training facility, since that was Fred's career after law enforcement. Just like Sheriff Ruben Schade, who took his place in 1940, J. F. Hartle was a conservative Christian with ties to Jackson, Cape, and Sikeston's MIA. Ex-Sheriff Hartle was also a dedicated area Freemason and knew many well-connected people in the community; just days after the crash he met some Cape legal eagles and the Cape County Medical Examiner down in Sikeston, ostensibly for a Selective Service Board meeting. Was the Cape County Medical Examiner, Dr. J. H. Cochran, someone who had just been called in earlier in the week to pore over the creepy alien bodies recovered from the crash? It seems quite conceivable. Were the bodies still in Sikeston at MIA airport headquarters? That does not seem likely. Was the Sikeston meeting afterwards a good opportunity for networking by those involved in MO41's first response? Anything's possible. . . . Another, similar meeting with Cape dignitaries and Sikeston officials was held not long after the crash as well. According to a 5/1/41 *Missourian* article, "Peace Officers to Meet in Sikeston," Sheriff Ruben Schade and the FBI's G. B. Norris (via the St. Louis field office) were noteworthy among those law enforcement officers meeting to ostensibly discuss policing and investigations by the "SEMO Law Enforcement Officers Association." The same questions come up about what some of these men knew of MO41 and its aftermath. . . . The lowdown on Charles Root's engagement came from the society column in *The Southeast Missourian*, 4/8/41.

Root graduated from the University of South Dakota and after eight years of military service was up to his own command, especially after toiling for the Army in Washington D.C. Charles became a "Congressional Liaison" for the Army air forces at their headquarters; oddly enough contemporary Lt. Col. Richard Root was an Army congressional liaison during the past decade, before retirement. I wrote to Richard on Facebook but he did not reply. But all of Charles Root's personal background and military history indicates he was very trusted, experienced, and knowledgeable, and not to be taken lightly. . . . Information on Truman's 1906 training for Army duty in Cape Girardeau and other background him comes from various biographies and Wikipedia, plus internet sites. The current hilltop Cape National Guard armory site (moved to from the '41 armory locale) was near where I often played and explored as a child, at Arena Park, not far from my family home in the 1970s. . . . *The Southeast Missourian* article for 4/14/41 told of the fire department's response to a Cape grass blaze that Monday. It was entitled "Fire Truck, Car in Cape Crash" (on page one). . . . Just how windy was it that Spring of 1941 in Sikeston? Linda Wallace uncovered one historical quote about the local airport: so many dust storms in April and May whipped up that they left the sum of 30,000 yards of top soil "spread to a depth of two inches on the home field at a cost of $15,000" to remove, an enormous total in the depression era. No wonder former Cape County Sheriff J. F. Hartle was placed in charge of removing dirt there. . . . Missouri author Janet Lowe recently published an obscure book based mostly on the 1973 Piedmont, Missouri, UFO flap, which I remember well as a boy. Lowe covered other regional unexplained aerial phenomenon in *"Of Unknown Origin. Or Is It,"* where she simply seems to regurgitate what research Ryan Wood first published around 2001. Lowe added that she felt the Cape fire department has records that shows data from the crash, which I have not seen. Likewise are simple rundowns of MO41 in a few paragraphs, summarizing Ryan's original story coverage - with small errors - within the recent books "*Mysterious Missouri*" by Ross Malone; "*Paranormal Mississippi River*," by Charles Cassady, Jr; and "*Weird Missouri*" by James Strait. Plus there's a 2011 book by Missouri author Lee Prosser that I have not seen, entitled "*UFOs in Missouri*," by Schiffer Books, but it was not well reviewed nor did it apparently feature any new information. Frankly *none* of these aforementioned publications were any real research assistance here, but at least they have helped keep the general story alive. . . . Paranormal writer Stan Hernandez says he goes to UFO and ET abduction conferences around the country and alleges he ran into young Phillip Reynolds at one to learn his and Walter Reynolds' supposed MO41 personal story. This according to Stan's account within BeforeItsNews.com where much on MO41 is aired. Hernandez is also reprinted, curiously, in RingInsiderReport.com, largely a sports site. The specific Hernandez article referred to here is entitled "*Eye Witness To Crash Comes Forward*," from 11/23/2012. . . . . . One likely candidate to have shadowed and/or wiretapped fireman Walter Reynolds would be policeman Marshall Morton, the future chief who was in the Army tank corps in World War I and was accepted to the FBI Academy in '41. . . . Quintin Williams quitting the Cape fire department on Tuesday 4/15/41 was chronicled in small articles in *The Southeast Missourian* on April 17th and 18th. An article a week or so before mentioned his desire to attend a "Young Democrats Club" on 4/18/41, in another Missouri city, but that would have been scheduled well in advance and not likely the cause of someone quitting their job so quickly that the newspaper was caught unaware as the story began to reveal itself. The fire chief, Carl Lewis, accepted the resignation, and needed a new fireman pretty quick to replace Williams. Lewis selected a Cape Girardeau policeman in the building, a man with experience with the local fire department less than a year before, patrol officer Roy Haman (not one of the two cops named in the Broadway accident earlier in the day). Roy simply moved from one room of the building to another; he might have been related to local camera buff Jim Haman, who worked with Frony and Chester Kassel. Anyway, this job move left Chief Barenkamp one man short; his police staff would be augmented by Mayor W. H. Statler who would soon make the selection of a new cop, according to the articles. . . . Fire chief Lewis retired in 1974, after 33 years total of service. It's difficult to conceive that even if he didn't go to the crash scene that night, he didn't find out about the event later on. . . . Ryan Wood provided the information discovered that Reverend Huffman wasn't hired full time at Red Star until 7/31/41. . . . An article in the newspaper indicated a couple of weeks after MO41 the Cape Fire Department was given new uniforms, but these were likely ordered and created before the crash incident, and were not "rewards for cooperation." . . . *"The Secret,"* the fine SyFy Channel special, when repeatedly aired resulted in very inquisitive, demanding viewers calling *The Southeast Missourian* newspaper headquarters, according to the paper's historian, Sharon Sanders. Sharon - who knew and worked with Garland Fronabarger and Ruben Schade - has had to field many a request for "the hidden file" or "the squelched story" that they feel the paper's owners and editors has been sitting on for decades. As far as Sharon knows, there is simply no blockbuster account of the MO41 crash stashed away in the archives, safe, or floorboards of the newspaper offices. She'd appreciate it if people would stop calling on the issue.

CHAPTER TWO: The same three main sources for Charlette Huffman Mann's recounting of the dead alien photo are again utilized: Ryan Wood, Linda Wallace, Linda

Moulton Howe all interviewed her thoughtfully and brought out the facts for their resulting sites and stories. Plus she responded through Ryan to some questions I asked in January 2015. . . . Perhaps one of the top places for any photographer at the MO41 scene *to avoid* afterwards would have been "Unnerstall Drugs," located on Broadway in Cape. That's because it was a family business related to local policeman Vincent Unnerstall. . . . Much of the same sources as mentioned earlier on Garland Fronabarger were used, including his *Southeast Missourian* obituary, plus I gathered data from my June, 2013, telephone interview with his son, John, living in retirement in Arizona, just outside Phoenix. Background on "One Shot Frony" comes partly from *The Southeast Missourian* online information, with recollections from those who knew him, including ex-reporter Sally Owen, who interviewed her co-worker repeatedly about his career. Sally and her husband Ray (who died in early 2015) both worked at the newspaper, lived across the street from my father in Cape, and I attempted to contact them for comment on G. D. Fronabarger and MO41. Sally was working at Southeast Missouri Hospital in 2014 when she replied by e-mail saying she tried to look up hospital records - at my request - for any files from the spring of 1941, but - as usual in this investigation - found nothing. "Record keeping in 1941 was just not as meticulous as it is today," she replied, describing any potential emergency service files as paperwork "long gone." . . . *The Southeast Missourian* also on occasion employed the photographic services of the two Kassel brothers, according to paper historian Sharon Sanders. The duo were apparently Chester Kassel's young adult sons, mostly studio portrait photographers in town, but at times they took pictures of people and places like G.D. Fronabarger. Southeast Missouri State University's Kent Library has possession of a large collection of Kassel photos, but unfortunately as of 2014 they are evidently in disarray at present. . . . Did G. D. Fronabarger have a fixation with air travel? A source who knew him later asserted that Frony "took great interest in the airport and gave it lots of publicity." The airport site would not have been far from the MO41 scene, according to some claims. Descriptions in this chapter about G. D. Fronabarger, and the previous chapter's reference to a local hearse sending a first response team, derive from articles in *The Southeast Missourian* special 1993 "Progress Edition." . . . The U.S. Army's *"Special Operations Manual 1-01"* from April 1954 was provided by Ryan S. Wood, authenticated after much testing by himself and his father, Robert. It certainly appears to be the real deal, a shocking and revealing "eyes only" government document that blows the lid off the recovered alien wreckage story. . . . To spell out the Huffman's family data for other interested researchers: Reverend William Guy Huffman was born in Lebanon, Missouri, situated in the middle of the state of Missouri, in 1888. Younger Floy M. Peters was born in 1897 and married Pastor Huffman in January of 1912. Their first son William Guy Huffman, Jr., was born 6/15/ 1916. He died on 3/7/74. Their second son, G. Wayne Huffman, was born on 7/13/1918. He died as an Army sergeant in Europe, bravely helping to fight the Nazis in World War II, and was buried in England on 1/28/44. Young Wayne evidently did not marry, and was a generous contributor of time and money to the Red Star organization and its Baptist affiliation with Cape's "Teacher's College," which later became Southeast Missouri State University. Obviously grieving Pastor Huffman and his wife were heartbroken, and they left Cape Girardeau later that year, for Oklahoma. On 9/15/59, the revered reverend died, and was buried in Lebanon Cemetery, in Laclede County, Missouri (which I visited in 2014). Floy apparently never remarried and was buried next to William after she passed away in Texas, on 9/16/84. If she had died the day before, it would have been on the exact day of the 25th anniversary since husband W.G. Huffman's demise. The couple's classy, gray, dual tombstone is still there to this day, also displaying their two sons' dates of birth and death. Floy left behind two brothers, but James Peters also died in '84, and Ed Peters died in 1997; no word on if they ever were told the MO41 saga. . . . The Army *"SOM"* mentions Kirtland AFB, tellingly, as a site for taking extraterrestrial discoveries, among other places. . . . With many of the startling details on Walter Fisk I defer to Linda L. Wallace, who also interviewed the dodgy man's kin, plus Ryan Wood and Stan Friedman, who both managed to go to Fisk's Albuquerque home and talk to this frustrating, evasive source in person. I discovered Fisk had died when I found his 2012 obituary online, a few months after the fact, and passed it along to unaware Ryan. Attempts to get FBI "Freedom of Information" results on both Fisk (by Ryan) and Fronabarger (by myself) have not proven fruitful; the FBI does not care to divulge such matters after initially checking Central Records Systems and claiming the two names did not turn up. The FBI did mention to me in a reply letter that more information on such subjects exists in its files, but "Congress excluded three discreet categories of law enforcement and national security records from requirements of FOIA. This response is limited to those records that are subject to the requirements of FOIA." Thus there could be much more there, but the U.S. government has found a sneaky way of ducking the public's desire to know. . . . Author Tim Good came up with a source from the Sandia/Manzano set-up near Kirtland AFB, a military man who wrote to him to say: "They told me that Sandia has examined *several UFOs* during the last twenty years" in this secretive government locale. The Sandia/Manzano situation is rivaled only by Wright Field in Ohio as a probable scientific study site for the MO41 materials. . . . Marshall Morton worked as Cape's chief of police from late 1941 to 1946, and during those years also took classes at the FBI training academy in Quantico, Virginia, along at times with Fritz Schneider, indicating the two men's closeness. They both graduated from the Quantico FBI Academy, and yet never officially joined the FBI, leading me to suspect they were utilized as FBI "assets" for information and investigations in the years afterwards. Both men were friends on the Cape force of course back in April of '41, and before that. Both lived in homes not far from police headquarters. I suspect Morton was a local Freemason, and certainly Schneider *was*. In 1953/'54 they were named co-chiefs of police in Cape, around the time Walter Fisk was fascinated in Kansas with the Huffman family photo. {Around '53-'54 the old Cape Girardeau fire chief, Carl Lewis, also returned to his top post, oddly enough.} Morton, died (childless) in 1974. Ex-Chief Schneider passed in 1970. MIA Captain Root died in 1998. Fire Chief Lewis left this world (also childless) in 1983. City Attorney Oliver in 1971. Mayor Statler died at age 52 of a heart seizure in 1962. Ex-Sheriff Hartle died in 1976, and MO41 Sheriff Schade passed on New Year's Day in 1986, while his brother Ben died in 1974 at home, at age 64. No record of them uttering a peep about MO41 can be found, although certainly the sheriff's brother Clarence did, briefly, as mentioned. Clarence died in 2008. The MO41 is now so long ago that not only have the original participants died, their *children* have in some cases also passed on, making the investigation even harder. . . . Perhaps just a coincidence, but both Mayor Statler and Sheriff Schade traveled from Cape Girardeau to Kansas City, Missouri, for important national meetings, within two weeks of MO41. Not every mayor in the country could be there, only a fraction, really, but "Hink" was for some reason among those invited and attended. The same could be said for Ruben and the conference of top sheriffs, held in K.C., both stories according to spring 1941 *Missourian* articles. . . . Ruben Schade, by the way, was evidently so fond of fellow conservative F. C. Donnell, the '41 Missouri governor and later senator, he named one of his two sons Dewey Donnell Schade. . . . It's trivial, but the timing is noteworthy: the former sheriff of Scott County, Mr. F. K. Sneed, died around the time of MO41, but supposedly while out in California, according to a Cape newspaper article. . . . Linda L. Wallace uncovered media reports on the secret insurance spy program, utilizing a 9/22/2002, *Los Angeles Times* article on it, penned by staff writer Mark Fritz. . . . Wikipedia also contributed some facts on these topics, especially on the complicated Manhattan Project.

CHAPTER THREE: Almost all of the *Topix* information explains itself and can likely still be found on those internet posting boards, under forums for Cape Girardeau and Scott County. Other comments could well be clues, or purposeful disinformation by those seeking to gum up the works, or just honest mistakes. . . . Scott County, Missouri, was named after local 1820s politician John Scott, considered by many to be the father of Scott County and the later small town of Scott City. . . . Scott County was almost entirely an agricultural territory, with small farm villages here and there, very little other industry. In the 1930s, one fourth of America's population was agrarian-based, each farm requiring great swaths of property for crops and/or livestock, so that logic dictates a spacecraft crashing to earth - at least in the United States - was fairly likely to have landed on a farm. . . . An *"Out of the Past"* article in a more contemporary edition of *The Southeast Missourian* related how in 1938 members of the Cape Girardeau Fire Department traveled down to nearby Scott County and held a special program for local citizens and volunteer forces in how to fight fires and handle other emergencies. Among those who showed up to instruct the small town and farm area denizens was said to be none other than Carl Lewis, soon to be promoted as the Cape fire chief, in charge of the department as of April of 1941. Thus it is more believable than ever that the Cape

Girardeau fire department would care about and respond to an emergency fire call in another, neighboring county, since it had people living there known to be helpful and interested in responding to - and perhaps reporting - area calamities. . . . What the elderly farmer source for "AllSouls" supposedly found in a Missouri farm field was described as unable to be creased or cut, with odd, inexplicable markings. This is precisely what rancher Mac Brazel said he found on his desert property near Roswell, New Mexico, in July of 1947, according to all reports of that world-famous crash tale, and by Jesse Marcel, Jr. (recently deceased), who was given some of the UFO debris by his Army officer father. It's also an exact description of what Wright Field secretary June Crain said was brought into her office one day, for her to toy with. However, in the past few decades these reports have been well documented in the national media, and thus someone looking to get attention could simply parrot these claims in his own allegation of privately holding onto - and testing - metallic alien ship remnants. . . . There was a small vehicular crash in Scott County that also needs to be mentioned, one from the Saturday night of 4/12/41. But it was not a UFO crash. According to the *Missourian* front page article a few days later, an Army sergeant named Victor Schoen died when his vehicle crashed on Highway 61 in Cape Girardeau, just south of Bloomfield Road, at 12:20 a.m., which is actually the immediate Sunday morning hours after that apparently-tumultuous night. Where was soldier Victor going at that hour, on a road in Cape while headed south to Scott County? The article explains Vic was headed to Ancell, which was a tiny town not too far from the UFO accident locale, but this was supposedly the young man's family home. Victor was allegedly on leave for the weekend and had been driving much of the day, away from his base at Fort Knox, Kentucky, which ironically was the site of General G. C. Marshall's visit just days earlier! Marshall left for other army camp inspections after he may well have been chauffeured around by Sgt. Schoen, whose main job at the base, according to the article, was as a "driver." Such an ironic final days. Vic Schoen's body was taken care of by "three Army officers" who arrived from Ft. Knox on Sunday to handle the dead man's remains and effects, in accordance to Army regulations, the *Missourian* reported dutifully. Also, down in Scott County the newspaper told of a second, unrelated auto accident on that Saturday night, 4/12/41, injuring three area men, on Highway 61 north of Benton, attended to by a Missouri State Highway patrol officer. When his son - later a Cape County Prosecutor - was contacted in 2014 he professed no knowledge of any alien crash in family lore. . . . An odd-sounding but memorable further *Topix* posting arrived in the fall of 2013, this one from an unknown Missouri farm community resident, possibly mixing two incidents in one: MO41 and the 1938 meteor that flashed through the skies in the same general vicinity. Or just describing one as best as possible, but it's likely not our UFO crash at all. A simple theory was formed back when, passed along by the grandfather: "A star had come loose and just came over and nearly crashing in the trees, lighting up the sky." Once again this sounds like a natural occurrence, focusing on the phrase "*nearly* crashing." It also happened "in the fall and not spring." So that cinches a fascinating similar-sounding memory but unfortunately not quite filling our bill. Linda Wallace is also a source of information on the area meteorite sightings and other events, her family from Sikeston, with a grandfather of hers running a restaurant there that often served PAC/MIA and airport personnel. . . . An anonymous *Topix* source posting a comment in 2013 on YouTube claimed that he was once a Cape Girardeau resident and that his grandparents were still living there, often mentioning "the Cape UFO crash." I attempted to make contact with the poster about his family's knowledge of MO41. Naturally this particular video was removed for copyright infringement, and with it, all the comments previously established. Another enticing lead that cannot be followed up on. . . . Evidently even for decades after the incident on their former farm, the grandson "AllSouls" wrote on *Topix*, the once-landowning and farming couple (his grandparents) would barely breathe a word of the crash, seemingly mostly out of fear of repercussions and continuing loyalty to America's security state. "Think about when this occurred," "AllSouls" reminded online readers. "There was a sense of patriotism and allegiance to our country that is diluted and devalued" nowadays, adding that in 1941 "people trusted and respected the government for the most part." I think that puts things into perspective nicely. . . . *Topix* featured many postings on the "UFO Crash Near Cape" threads from a man named "Cousin Ed." He initially claimed that the crash site was near the new Wal Mart store off Gordonville Road, "near the gas station." Then Ed moved the site a few miles down the same road, to near the new Notre Dame high school, "south side of the road." When pressed for more info, Ed claimed he lived in fear to tell more details and generally clammed up on the true location, but continued to express his belief that the 1941 accident was genuinely extraterrestrial and suppressed by the U.S. government.

CHAPTER FOUR: Linda Wallace's kindle book *"Covert Retrieval"* was a fascinating read of her family history, but frustratingly with no names or helpful identities attached, apparently to protect privacy. Thus I had to make up some. Linda informed me she is not sure if her kindle book will ever be published and released in hardback or paperback form. . . . Still not technically in the U.S. Army, Linda Wallace's father worked under the command of base Captain Charles B. Root at the Sikeston airport. By 3/20/41, there were a whopping 101 new students readying for the next set of classes at the airport, stated in Cape's newspaper. Also, Captain Root enlisted at least a dozen MIA cadets to dress in full uniform and act as an "Honor Guard" at his 5/31/41 wedding in Sikeston, according to the 6/2/41 edition of *The Southeast Missourian*. Root married a local woman who worked as a school teacher in town for the previous four years. They left the ceremony for a "month-long honeymoon," according to the article. The commander of the airfield operation was allowed to just walk away halfway through the ten-week training course, which began 5/2/41. The newlyweds didn't have long to settle into the "Park Avenue Apartments" in Sikeston, as the news story related, as a few months later Captain Root was transferred to an airfield in Texas. Presumably his new wife Camille went with him. Charles went on to serve his nation honorably in the coming years, moving from base to base, airfield to airfield, office to office, especially during the war years, according to the online USAAF military biography. . . . Linda Wallace's story of what "Sam" recalled of his "Grandma Judy" recollecting the "day after" the crash - i.e., it being loaded onto a flatbed truck by uniformed military men - seems to further negate or erode confidence in the Phillip Reynolds, Jr., claim. That is, Phil's grandfather's alleged confession, that he stuck around the crash site to watch the Army load the damaged space vehicle (and then was booted out for swiping debris). Walter Reynolds was the fireman alleged to have been called to the crash site that night, and thus it is most unlikely he was "still there" when the evidence was picked up by the military the next day and hauled away, if it was indeed done in that manner. However if Walter witnessed only bits of debris being picked up, not the ship itself, the story makes sense. Judy's allegation that the Army came in the next afternoon is the more believable tale of how the evidence was handled, since it would take a while for the story to work its way up the military chain of command and for decisions then to be made on just *how* and *when* to recover the larger parts of it as quietly as possible. That process would take at least half a day, one would think, especially if it was felt by the Army brass that the once-flying object would be heavy - which by all accounts it turned out not to be. . . . Ryan S. Wood pinpointed two other figures possibly involved in the two retrievals: the Sikeston MIA Intelligence officer, James Lightle, and the 309th FTD base medical corpsman, Louis Digiglia. They do not show up in the USAAF military biographies online research site. If the twosome were around at the time of the MO41 crash, they would have most likely been an integral part of the two-day recovery process. . . . The MIA training headquarters in Sikeston, at the airport, was located at 816 N. Kings Highway. It was closed long ago, but the airport is still there. . . . The East St. Louis PAC at one point actually produced its own promotional newspaper, but many period pieces I utilized herein - about the Sikeston PAC and MIA - came from *The Sikeston Herald* of the time and were reprinted online within Ancestry.com. Information in that site was provided in 2005 by Linda Wallace and her contact Carol Bowman. . . . A 5/22/41 *Sikeston Herald* newspaper article confirms for us that men in and out of uniform ran the Missouri Institute of Aeronautics at the local airport. Showing photos of men employed at the MIA facility, the sub-headline read: *"Civilian Administrators Dressed in Business Attire, Military Personnel Dressed in Uniforms."* Thus we have a small but significant clue about "Gramma Judy's" recall of the recovery team at the daytime crash site, including as it did "men in uniforms" with one man "in a suit" paying her father to heft the crashed objects. Linda stated that many men who weren't technically in the Army at the Sikeston MIA, including her own father at first, were encouraged to wear uniforms at times, to help make things more confusing then and especially now. . . . Oliver L. Parks might have been called the night of the MO41 crash and informed, then hopped the first flight down to Sikeston the next morning to investigate, although there is no proof of this. He loved aircrafts and flight stories, that was nearly his entire life, and *this* was a whopper he could not miss out on, it would seem. . . . Wright-Patterson Air Force Base near Dayton, Ohio, was not home to an Oliver Parks PAC, and no, Mr. Parks was not in any way a part of the "Park College" that sits right in the midst of the W-P campus. This particular college is very old, going back centuries now (see Wikipedia for details). . . . Information on Oliver Parks and Henry "Hap" Arnold can also be found online, particularly Wikipedia and other biographies. Arnold and his "Project RAND" proposal for space flight came from a web page called *"Deep Secrets"* with material penned in 2009 by Anthony Bragalia. Wildly, atom bomb manager Gen. Leslie Groves also joined RAND. There is plenty more to the UFO connections and technology within RAND, but time and space here are limited. Henry Arnold died in 1950, having a tenuous coronary condition for years. Oliver Parks died at age 85 on this author's birthday in 1985. . . . Linda originally featured on her website a strange Charleston, Missouri, news story from 4/6/41 just days before MO41. It regarded a "negro" who was a local farm employee who claimed he saw a submarine pop up out of the nearby Mississippi River, then went aboard it, outside Charleston that previous weekend. A later follow-up article revealed the African-American farmer had admitted that he made up the story, but it is a revealing tale for us in so far as it shows that before the UFO crash, Missouri State Highway Patrol officers and FBI agents (and even the Coast Guard!) showed up at the man's farm and interviewed the claimant, the report claimed. Plus, the area newspaper took the story seriously enough to run an article on it, then another. According to the odd tale, the FBI agents involved took the farmer north, to the St. Louis field office for more intensive questioning, bypassing Cape Girardeau's closer, new set-up, which may be where these original agents were situated in the first place. The matter

would have been dumped into the lap of Agent in Charge
G. B. Norris and his sidekick, John R. Bush, at their St.
Louis office, and they in turn reported to Assistant Director
Clyde Tolson and Director J. Edgar Hoover in D.C. It may
have been just days later that these same southern Missouri
law enforcement officers sped to the MO41 scene and then
reported back to their superiors on the amazing case. . . . To
me and also to Linda L. Wallace it is surprising and
inexplicable that the well-run, well-staffed, and successful
Parks Air College in Cahokia was rather abruptly closed in
August of 1943, and the Sikeston MIA training school was
similarly shut down, in October of 1944, all while the
second world war was still underway and well-trained
pilots were in demand. Linda reports that during the war,
Mr. Parks allowed new management to take over the day-
to-day operations while he still maintained other
government contracts and businesses elsewhere. Yet
according to newspaper articles, Linda indicates, Parks
continued to visit the Sikeston MIA operation. "I could not
determine the exact reason for the change," Linda reported
in 2014. It may have been felt by the Army and the air
forces that other strictly military airfields would provide
better training facilities for teaching cadets, rather than
civilian airports and operations going on at the Cahokia and
Sikeston sites. To make matters more complicated, the
new management started reworking old records, and may
have just dismissed outdated files that might have proven
helpful to this day. Even more confounding than that, a
large fire in St. Louis destroyed many U.S. Army records.
Oliver Parks also opened - with other investors - a flight
training program at, of all places, *the Cape Girardeau
airport* - then called Harris Army Air Field, not really that
far from the MO41 countryside crash site. Ironically it
bore the initials "CIA" before there was a Central
Intelligence Agency. The Cape Institute of Aeronautics did
not last very long either, but it managed to train the
USAAF's 73rd Flying Training Detachment. It closed

during the summer of 1944, even before the PAC/MIA operation in Sikeston. Apparently new management wanted to reorganize and clean house, perhaps in conjunction with changing American wartime government requests and requirements. . . . The Cape Institute of Aeronautics was not the first training school at the Highway 74 airport, as crude and simple as it was in the early 1940s. *The Southeast Missouri* on 3/18/41, chronicled the opening of the "Consolidated School of Aviation." The training school didn't last that long in the scheme of things, but was likely the top attraction in town to young Rush H. Limbaugh Jr., the aviation-obsessed father of the now-famous radioman. . . . In a rather odd coincidence, there was another crash in the spring of 1941 that brought together the MIA and another weird angle of MO41. It seems that *The Sikeston Herald* reported on Thursday, 5/15/41, that the MIA sent 26 cadets to Lafayette, Indiana, to attend the funeral of a cadet who recently lost his life in a tragic accident when two airplanes collided on the ground at the Sikeston airfield. Lafayette is the home of Purdue University, where a MUFON report claims crashed MO41 alien airship engine parts were at one point allegedly brought in by government representatives for a private study by post-graduate students (mentioned in *"Hangar One"* on the History Channel). Linda's father was not amongst those training cadets who attended the funeral, according to the article, and the young fliers who did were almost entirely new recruits, likely unfamiliar with the UFO crash the previous mid-April. The young pilot who died in the accident had attended Purdue for two years prior to his air schooling in Sikeston. Ancestry.com was helpful in calling up this old newspaper information in online research. . . . Linda said in 2015 her mother died "a few years ago." Author Wallace's family source, plus a second "senior source," both mentioned the crash site for MO41 as in the "Benton" area, many miles north of Sikeston and many such miles south of Cape Girardeau. It was near a farm with a specific family name Linda did not reveal, and Benton *was* a "crossroads community" as Linda mentioned others' describing as a clue in those days, but. this general site seems awfully far away for a resident to have called the Cape Girardeau fire/police departments, and for the Cape County Sheriff to have also arrived on the scene, since as mentioned Benton was Scott County's seat of government. Still. who knows? Until the crash site is revealed firmly, all possibilities are on the table.

CHAPTER FIVE: Linda Wallace's website "http://www.seekingmoinfo.com/" is a good source for the Turner Holt revelations, featuring an interview with the Holt sisters she helped record in 2008. . . . Linda was friends with the late UFO investigator and author Kenny Young, an Ohio-based researcher who once wrote a book about his experiences in digging into paranormal aerial phenomena. He even wrote a chapter (unseen by this author) on the 1941 Missouri crash and recovery, and briefly assisted Ryan S. Wood in his own MO41 investigation. Sadly, Kenny died of leukemia in 2005, at age 38. . . . Turner Hamilton Holt was born 3/24/1894 and died on 2/5/1960, just five months after Reverend William G. Huffman, ironically. Turner's sole wife, Mrs. Vina May Clark Holt, was born on 1/23/1901 and died in September of 1993. They are buried side-by-side in a Shenandoah, Ohio, cemetery. In addition to two Holt daughters mentioned - Lucille, born in 1921 (died in 2009), and Allene, born in 1923, they also had a third daughter, Geneva Marie Holt (born in 1922), who doesn't seem to be interested in the ET allegation, or at least speaking publicly about it. Turner H. Holt also had two brothers and two sisters, all long since deceased. Pastor Holt's granddaughter, Eloise May Gramly Watson (born in 1947), offspring of Allene, posted their basic data online and expressed the desire to learn more about her family heritage, her somewhat famous grandfather, and his riveting Capitol Building UFO claim. Turner and Vina May also had two other grandchildren and a great grandson. . . . Investigator William "Bill" Jones also dove into the Holt sisters' story and its ties to Cordell Hull, and wrote expertly (along with co-author Eloise Watson) about the fruits of these efforts in *"Pre-World War II Creature Retrieval?"* The article appeared in the Winter 2001-2002 issue of "International UFO Reporter." . . . Grant Cameron undertook Holt research, then aired online background material and an April 2009 interview with the Holt sisters as well at JerryPippin.com. His recorded material contained the "three glass jars" exclamation. . . . A few other online UFO sites air the Hull-Holt claim too. Cordell Hull's life story and profile can be found on Wikipedia and in biographies and history books, and it should be stated that he appears to have been a wise, competent, hardworking, and trustworthy man, having truly earned his high status of expert statesman in the Roosevelt administration, although as mentioned elsewhere he wasn't always on the president's wavelength. . . . . Turner Holt's 1956 book, *"Life's Convictions,"* is out of print and pretty rare, and I have not found or read it. It was published by Vantage Press, 95 pages. . . . CBS News reporter/commentator Bob Schieffer issued on the air a story of President John Adams wanting the large sub-basement storage area utilized for the American Founding Fathers. Schieffer spoke extemporaneously during the live CBS coverage of President Barack Obama's January 2013 re-inauguration; he revealed how the Capitol Building's historian told him stories of the uses for the lower parts of the famed building. As Obama and his administration and honored guests were shown descending flights of narrow stairs, leading down to the Capitol Building's exterior balcony platform for the inaugural ceremonies - overlooking still more stories below - one could more easily see how there really would be a *great* deal of offices and storage space in the sub-basement levels. The Capitol site is obviously truly enormous and well-designed, with solid construction to hold all the people assembled for daily activities and visitors. . . . The scoop on 7/1/40 additions to the Capitol Building's police force come from the fine research of Secret Service expert, Vincent Palamara, an online friend. . . . Roosevelt's blood anemia crisis that Spring of '41 can be found in the pages of *"FDR's Deadly Secret,"* 2010, by Eric Fettman and Dr. Steven Lomazow, who also produced a fine website on the subject, within scribd.com. . . . Information on those who laid in state in the Capitol and also on FDR's funeral arrangements were mostly from Wikipedia, and a bit from the PBS special on *"The Roosevelts,"* from September 2014. . . . I did some research into large glass jars manufactured in East St. Louis area after learning that Linda Wallace had a source who claimed that a PAC member was given orders to fly from Cahokia down to Sikeston to "handle the bodies." Was it even possible this obedient military man could have picked out three glass jars and flew them to the Sikeston MIA site at their airfield? In checking I discovered that some glass manufacturers did create such large containers in nearby Alton, Illinois, and that if it was explained to management in April of '41 that it was an emergency, such a person *could* have conceivably bought them at the plant and hauled them away for a top secret military mission. Records from this 1941 glass production plant are nowadays gone, the owner told me through e-mails, but it was at least possible in 1941. Since the Army handled the crash and the bodies one would naturally assume the most logical place for the glass "jarring" of the aliens would take place in private at Walter Reed Army Hospital or Bethesda Naval Hospital. Background information on those famous military medical facilities, as well as Fort Meyer, near Arlington Cemetery, can be found on Wikipedia and their

individual official online sites. . . . Information on General
Marshall is likewise found online, and in a fine 2013 book
called *"Roosevelt's Centurions,"* by Joseph E. Persico.
That author describes the rather dour, somber General G. C.
Marshall living at Fort Meyer and often roaming the
grounds of Arlington on foot or on horseback. Indeed
Marshall would be buried there in October of '59, the
funeral service at Fort Meyer attended by presidents
Eisenhower and Truman and many other dignitaries in
attendance. Marshall also had a home in Leesburg,
Virginia, but spent the vast majority of his time in the
Arlington/Washington area, working out of the War
Department building much of the time. GCM's beloved
Fort Meyer remains to this day the main headquarters for
armed service personnel in the National Capitol Region,
built in 1861. . . . Art Bell, George Noury, and their
frequent guest and friend Linda Moulton Howe can be
heard on their respective radio shows and on their internet
web sites. Linda laudably produced a web page within
Earthfiles.com that airs a thoughtful interview she did with
her friend Ryan Wood in 2003, entitled *"Majestic Twelve
Documents: Extraterrestrial Technologies,"* which proved
helpful. Linda's 1/10/99, interview (re-aired in early
March '99) with Art Bell on *"Coast to Coast"* may have
been the first big national airing of MO41, at least to those
interested in the paranormal. Linda was evidently
exploring much of the information first published in *The
Southeast Missourian* by Peggy O'Farrel, six months
earlier, plus her own intrepid investigation, including her
personal interview with Charlette Mann. I overheard LMH
mention "Cape Girardeau" while turning the radio dial that
March night and pumped up the volume. It was the first I'd
ever heard on the case. I hurriedly grabbed a pen and paper
and scribbled some notes, then typed them up, and sent
copies to some family members before tucking my original
away. I did not find this 1999 page again until the summer
of 2014. . . . In my LMH radio notes I see that she

mentioned how German scientist Krause (who apparently died in 1990) once stated he was also informed by *his* source, a man from "Majestic Twelve," that President Truman formed that top secret ET research group in 1947 following the Roswell incident, in order to run an end-around the FBI and congressional red tape. . . . *"The Secret"* in DVD form can now be found for sale online, within Ryan Wood's fine, must-see website, majesticdocuments.com. *"The Secret"* originally aired on the SyFy Channel (then called "Sci-Fi") in the spring and summer of 2003, repeated a few times since. Ryan's 2003 downloadable .PDF exploratory paper on MO41, called *"The First Roswell,"* augmented interest in the innovative TV program. It is a fine document, still viewable online, and very much like his MUFON article, but lists "north of Cape off Highway 61" as a possible crash site. Not likely accurate in my opinion, but this was back in the very early days of the investigation, and more facts have been uncovered since. MUFON's journal with Wood's report was originally published in their November 2001 as Issue #403 and can still be read online. . . . Ryan Wood partial quotes on the remote viewer he hired and the 1941 aerial photography he discovered come from a 2002 interview he did with Linda Moulton Howe, reprinted on her Earthfiles.com website.

CHAPTER SIX: The FDR and Marshall memorandums (and the Bush-to-Truman letter) explored herein were provided by Ryan Wood, and can be found on his marvelous web site MajesticDocuments.com, and in his aforementioned book. . . . Controversial opinions and insights from "Jerome" herein are certainly not documented within government stationary from someone with a discernable and check-able intel background. However, the anonymous spy source's information sounds sensible and logical regarding recovered hardware like MO41, via the MAJESTIC and JEHOVAH program claims. This typed letter to Tim Cooper can again be found within Ryan Wood's mentioned document http://www.Majestic-Documents.com website. Wood and Cooper refer to Jerome as "Source S-1," describing his work aptly: "the memo is hard to summarize because of its varied content" but its points are "rich in detail" and "IMPORTANT." Jerome's letter starts to veer off into notions that border on sounding a bit paranoid in the final two pages, regarding "materializing aliens" and a possible future invasion of earth. I find much of his provided information not just compelling, but believable overall, however. . . . Dr. James Conant is a name from the past that needs to be investigated further; he likely knew a great deal about the MO41 affair and its scientific applications for America's defense programs. James was fairly close to Van Bush, and an accomplished academic, becoming President of Harvard University. James was a chairman of one important defense committee and served as Bush's Deputy Director of the Office of Scientific Research & Development, which officially began on 6/28/41. Bush, Conant, and the OSRD was "heavily reliant upon the involvement of the faculty at the Carnegie Institute," based out of Washington D.C., according to Linda Moulton Howe in her Earthfiles.com research. The main focus of this group's participation was further centered around the esteemed institute's Department of Terrestrial Magnetism, possibly indicating once again an atomic/magnetic device was the source of the propulsion unit on the alien spacecraft. The folks at OSRD, Carnegie, and the DTM are the most likely regular research resources and/or actual staff for President Roosevelt's "Non-Terrestrial Science and Technology Committee" and/or the scientists responsible for exploring and copying the extremely classified atomic engine of the crashed ET ship culled from Missouri. The OSRD was largely responsible for the development of radar and radio proximity fuses, along with the atomic bomb program, as examples of its exemplary devoted defense work. Some of this information was provided in a report quoted by LMH and originally from - of all sources - Sandia National Labs, further indicating a real connection there, one that might even attach Walter Fisk, who knows? . . . After some research, a Canadian governmental employee, Wilbert Smith, wrote a colleague in a now-famous 1950 memo (uncovered decades later) on recovered UFOs: "The matter is the most highly classified subject in the U.S. government, rating higher than the H-bomb. . Flying saucers exist. . A concentrated effort is being made {to understand the recovered crafts} by a small group headed *by Dr. Vannevar Bush.*" This indicates Bush was likely still researching the MO41 vehicle (and others?) for President Truman, up to *nine years* after the Cape Girardeau crash. . . . Dr. Bush had two sons who could not be interviewed in time on MO41; both died around the turn of the twenty-first century. Van Bush's "Office of Scientific Research & Development" was set within FDR's "Office For Emergency Management Of The Executive Office Of The President" (what a mouthful!). This shows us again his direct line to Roosevelt, and reinforces the urgency for the atomic bomb project and other scientific advances that were being utilized in defense, infused that spring and summer with MO41 technology, I feel. . . . The 10/9/41 Van Bush meeting with FDR and his VP was mentioned on page 218 in *"Oppenheimer,"* a 2005 bio of the famous physicist. Also included therein is the Roosevelt secrecy directive on the atom bomb development project. The Bush-FDR/Wallace meeting is confirmed in his online daily diary, taking place at the White House at 11:30 a.m. . . . Info on spy-master William Stephenson comes from *"The Irregulars* and its author Jennet Conant. . . . Information on Gen. William J. Donovan (an old law school chum of FDR's) and his intel positions, more on Stephenson and Dahl comes from Wikipedia. As does the amazing lowdown on the *"Charlie and the Chocolate Factory"* book plots. A third book was planned, for a trilogy, but author Dahl only wrote one chapter of *"Charlie in the White House,"* where the fictional boy would have an adventure with the American president Gilligrass. . . . Stephenson died in 1989 and Dahl in 1990. But Donovan had a shorter life and an uglier fate; he eventually suffered vascular dementia and strange, mad hallucinations before he died in 1959. . . . It's interesting how Gen. Bill Donovan, Gen. George Marshall, Dr. Ross McIntyre, and Rev. William Huffman all died within months of each other in 1959. . . . James Roosevelt, by the way, was a Freemason and from 1955-'65 a U.S. congressman from California, working in the Capitol Building, of course. He died in 1991. James and "Wild Bill" Donovan worked so closely together it was James who called Bill to notify him of the Pearl Harbor attack. There is no specific evidence that James Roosevelt, easily the closest child to his father, was informed of MO41 but if I were a betting man I'd put money on it. Two of his children responded to my letters in late 2014 and stated they had no knowledge of MO14, but it was such a top secret it likely was not dared discussed, even in later years. *Time* magazine once suggested that James be dubbed the "Assistant President of the United States," he knew so much and had so much influence in the 1930's and early '40s, according to Wikipedia. . . . Also a close aide to FDR in the White House may have known about MO41 was 1941-to-1943 "Correspondence Secretary" Marvin McIntyre (no relation to FDR physician Dr. Ross McIntyre). Marvin may have been the one who took dictation on early FDR MO41 memos, but who manually typed and sent them out is unknown. Aide Daisy Suckley is a possibility, in that she might well have been present at the White House when the first calls came in on the 4/12/41 UFO event. . . . Some of the Harry Hopkins (1890-1946) info can be found online, in books, or specifically his Moscow trip within page 62 of

"Rendezvous with Destiny." At that time, on his cross-continental trip to Russia, Harry was allowed by channels to be referred to as a "special ambassador" to the president, and "permitted to cross frontier stations of the USSR without examination of luggage." Pretty posh treatment, whether he was a secret Soviet spy or not. . . . Reference to the secret Army airfield outside D.C. comes from the 1977 *Time-Life* book *"The Home Front: USA."* . . . In my opinion, Gen. Marshall has been appropriately honored for his dedicated coordination of MO41 materials by the U.S. Government: near Huntsville, Alabama, the nation's "civilian rocketry and spacecraft propulsion research center" is named "The George C. Marshall Space Flight Center." This base helped create Saturn launch vehicles, the space shuttle, and the International Space Station design and assembly. According to Wikipedia, much of this basis for these operations were born out of the 1945 "Operation Paperclip" program that started out in Texas and New Mexico, and was relocated to the new Alabama center as the late 1940s progressed. All army-related space programs were transferred to the Marshall Center, which was named after the late general-turned-Secretary of State in 1960. This was specifically approved of by then-President Eisenhower, who had previously worked under Marshall at times. It's conceivable Ike knew that Marshall was a key figure in guiding the Missouri discoveries and other IPU finds he handled into a secret space program partly operated by Paperclip and RAND, via GCM's friendship with Hap Arnold. Eisenhower could have named the space center after anyone or anything, but chose GCM due to MO41 and perhaps other UFO crash hardware technology management. Why *else* would George Marshall's name go on a *space flight* center? . . . Paperclip produced the German scientist Oberth, whose '54 quote used herein was originally part of an article in *"The American Weekly,"* 10/24/54, was elsewhere in the news then, and has popped up in many books since. His famous

repatriated ex-Nazi colleague, Dr. Werner Von Braun, also hinted publicly at recovered UFO technology, but was not as explicit, although his spokeswoman rather stunningly told the media years later "he knew about the extraterrestrial issue." (see AboveTopSecret.com for more) . . . Two key figures who worked under G. C. Marshall in 1941 likely knew and handled MO41 materials: Colonel Rufus Bratton (1892-1958), a top intelligence officer and aide to GCM; and the Assistant Chief of Staff for Intelligence, Brigadier General Sherman Miles (1882-1966). These two, along with Colonel Orlando Ward (1891-1872), Marshall's secretary, were amongst the closest and most trusted Army personnel GCM knew and relied on. If anyone had been ordered by Gen. Marshall to help covertly transfer the MO41 spaceship, shrapnel, and bodies, it would have been these three well-regarded officers. More needs to be learned about their whereabouts on that Easter weekend in 1941. It is quite plausible that all three were in touch with Senator Harry Truman, whose military committee work had officially begun on 3/1/41, at the Capitol and senate complex. . . . Oval Office visitor Vannevar Bush and his brilliant career can be found in some books, YouTube clips, and Wikipedia, along with of course the much more famous Albert Einstein, of course featured in various TV biographies, books, magazine articles, and documentaries. Van's Deputy Chairman of the OSRD from 1941 to 1945 was Dr. James Conant, a highly educated university professor who was also likely quite intimately familiar with the MO41 discoveries and their application in America's fledgling atomic propulsion and weapon systems. . . . Data on the secret bomb-financing meeting with Marshall, Stimson, and Bush briefing the three politicos at the Capitol came from page 291 of *"Truman,"* the Pulitzer Prize-winning book. . . . Information on the 1941 Carolina Maneuvers comes from Wikipedia and a few other online historical sources on those war games. More specific data utilized herein on the alleged UFO materials possibly found there (CAR41) comes from both Ryan Wood's helpful 2005 book, *"Majic: Eyes Only,"* and the website exposeUFOtruth.com. Wood cites UFO investigator Walter Webb's research on the mother's claims on her son Guy's revelations. . . . Information on Dr. Robert Wood's initial interest in UFOs comes from a Timothy Good account. . . . Facts on the OCD from FDR's executive action on 5/20/41 stem from Wikipedia and via the book *"No Ordinary Time"* by Doris Kearns Goodwin, 1994. She revealed that Eleanor Roosevelt got involved in the OCD and turned it towards social services as well, not Franklin's original intent. Identical resources were used for the wording of FDR's 5/27/41 radio speech. . . . The May 1941 quote by FDR on his lying comes from the "FDR Scandal Page" web page, and it was also utilized within the 1994 PBS *"American Experience"* special. . . . Information on and about Franklin D. Roosevelt and his big personal secrets listed herein have been the subject of books and TV productions, such as the special 2014 PBS television mini-series *"The Roosevelts: An Intimate History."*

CHAPTER SEVEN: Almost all of the information in this chapter has been sourced in the above other chapters and needs no further sourcing or explanation. However, I can provide some backup data herein. Such as: Tom Cantwheel reasonably claimed the Missouri (and Louisiana) spacecraft recoveries were controlled by the U.S. Army and studied intensely behind the scenes in the 1940s by American and imported German scientists. Their special designs and technologies were applied into a new "S" craft, a manmade disc mockup, a somewhat-flawed round spacecraft with human pilots that could achieve noteworthy speeds and heights in altitude, thanks to secretive testing at Wright Field in Ohio, then later at two test sites in New Mexico

(Kirtland AAF and Alamogordo AAF). This "reconstruction commenced in 1945," Cantwheel explained, but in the next two years the program struggled due to lack of funding, plus whenever the finished result was flown in great secrecy, pilots proved prone to illness and even death from decompression and radiation while inside the cobbled craft. This claim dovetails neatly with the '47 "White Hot Report" mention of just such a hybrid project. The top secret vehicular project was eventually scrapped altogether. It was aided, by the way, via the assistance of German rocket scientists smuggled out of Europe in the months after the end of WWII hostilities there, in what is now commonly known as "Operation Paperclip," according to retiree Cantwheel. . . . As MO41 gained momentum in the media an obscure West Virginia school teacher and mathematician named George Dudding tried to cash in, slapping together a "NON FICTION 32 page stapled book" (as he put it) regarding just the Charlette Mann story about her grandfather. This nearly unseen *pamphlet*, really, was released in a string of self-published mini-books by the paranormal-loving author on subjects ranging from monsters to "Mothman," ghosts to "alien shootouts." These largely overlooked efforts may not have received more than a handful of readers over the years, but *"The Cape Girardeau 1941 UFO Incident"* can at least still be found to this day for a few bucks on E-Bay, showing *some* lasting power. . . . Longtime movie and television actress Shirley MacLaine is well known for her belief in extraterrestrial visitation, and in her 2007 autobiography she reiterated her long-held understanding that the American government is lying to the public on the subject, willfully covering up a tremendous story of international importance. According to her own fervent research since the 1970s, alien crashes began as early as 1934(!), were all scooped up by the U.S. Army, and were stashed away for secret scientific studies much like MO41, of which Shirley informed me – through an aide - she has no actual knowledge. "All material and evidence related to UFO technology is located under research and development on the Foreign Technology desk" at Wright-Patterson, presumably, Shirley wrote confidently in '07, citing some military sources, plus her friend and UFO researcher of some renown, Dr. Michael Wolf, PhD. Wolf alleges he used to work *for the U.S. government, with friendly aliens* at a secret American government installation some decades ago. According to *his* sources, "The first UFO came down in 1941" (thus contradicting the actress). Dr. Wolf then goes on to spoil the MO41 implications of this statement by strangely claiming this was the year a disabled extraterrestrial spacecraft splashed down in the Pacific Ocean near San Diego and was recovered by the U.S. Navy. It might well be that either Shirley is a bit confused in relating the story in her biographical essay, or that Michael was a bit confused as to his 1941 otherworldly crash sagas. A UFO *may* well have crashed in the Pacific off the coast of southern California, but this most likely would have been what was reported within the George Marshall memo to Roosevelt in March of 1942, regarding the recent so-called "Battle of Los Angeles," where supposedly one spacecraft crashed in the San Bernardino Mountains, and another one allegedly ditched in the ocean that late February. . . . Prominent, often-skeptical UFOlogist Kevin Randle has also investigated many an ET crash-landing saga for his series of intriguing books; he dug into the MO41 affair within the pages of his comprehensive 2010 tome *"Crash: When UFOs Fall From The Sky"* and then again in *"Alien Mysteries, Conspiracies, and Cover-Ups,"* in 2013. While the cynical Randle did not give the MO41 affair great lengthy discussion or total endorsement, he came across as obviously intrigued by it, offering little if any real criticism of the case, unlike almost every other UFO incident examined in his publications. The skeptical, questioning Randle pointed out that to her credit, Charlette Mann never changed her original story, but only expanded it mildly over the years with a few extra minor details, and that she did *not* promote or support a controversial alien-theme photograph that Leonard Stringfield sent her in the 1990s. This unusual, eye-catching black-and-white picture - showing a small ET being led by men in trench coats and uniforms - is "an obvious and well-known hoax," which Miss Mann did not fall for, further aiding her case for authenticity. . . . In 2002 noted UFO author Richard Dolan mentioned MO41 in *his* comprehensive book, but again, not at great length and with notable enthusiasm. However he did not dismiss or pick apart the case, either, and discussed it openly and fairly on a national radio show. . . . Quotes about Ronald Reagan and Steven Spielberg's great interest in UFOs and ETs can be found in various topical books and websites. . . . Syndicated columnist and *"Haunted Missouri"* author Jason Offnut has written about the Cape Girardeau crash, back in 2006, and so did a Columbia, Missouri, newspaper reporter a few years later. In fact, there are several other such authors who have also passed along the basic rudimentary facts of our subject, too many to recount here. One example, however, should be noted. Ohio-based UFO researcher and writer Kenny S. Young enthusiastically dove into the MO41 narrative before he passed away in 2005, and helped uncover side stories related to other odd UFO sightings - and a downed weather balloon - in southeast Missouri in the early 1940s. Kenny was a dedicated investigator and would generously share info openly with others, such as Linda Wallace. *"Commotion Near Cape Girardeau"* was a key chapter in his 2008 posthumous book. Kenny's hard work over the years should not be forgotten, and his fascination with MO41 without finding fault within the curious case should be emphasized perhaps most of all. . . . In 2014, MUFON investigator Margie Kaye stated that she too had been researching the MO41 claim and would be in the Cape Girardeau area in the future. She startlingly feels MO41 is only "one of *three* UFO crashes in the 1940s" in Missouri! It will fascinating to someday read more about *that*.

BONUS CHAPTER: Associated Press reports at the time of President Roosevelt's death were often printed with no byline. I utilized some from *The Southeast Missourian* on 4/12/41; the 12th's special late edition; and Monday the 14th's newspaper. Plus a few tidbits in news stories from some online and book sources. One story curiously claimed FDR wore a blue suit with a vest, which he almost never wore, and a necktie instead of his more common bow tie, but it is clear that at the time of his death, Roosevelt was instead wearing a grey suit with a red necktie, even noted by Suckley in her diary and seen in Shoumatoff's painting. . . . Various online internet posting board comments were main sources for the startling FDR suicide claim utilized herein. Among these are Amazon.com, Answers.Yahoo.com, Snopes.com, Educationforum.com, and a collection of comments from lovcap.blogspot.com, which features additional blog stories under the title *"Lies Your Teacher Taught You."* There are evidently some more FDR opinions and comments under the link SteveMcQayle.com, but I could not get this outdated site to download. Then there was pertinent blog information on the site Ersjdamoo.wordpress.com. . . . Vice President Truman was scribbling a letter on 4/12/45, and at the bottom it shows he came back to quickly mention how his world had suddenly changed at the shocking news of FDR's death. "What a blow," he wrote, an enigmatic response to a supposed stroke by FDR, but a fitting revelatory statement if he had in fact heard instead of a traumatic suicidal gunshot to the head. . . . Daisy Suckley's diary with the entry on the president requesting a gun was read aloud in the 2005 History Channel special, *"FDR: A Presidency Revealed,"* which can be seen on YouTube

(Part 16). . . . The Kermit Roosevelt information was provided by the 2014 PBS show *"The Roosevelts,"* along with Wikipedia and some web sites on his life and death. Almost identical descriptions can be said of research on Lucy Mercer and Missy LeHand used herein. Noted author Doris Kearns Goodwin wrote in her bestseller *"No Ordinary Time"* about Mercer and Shoumatoff fleeing Warm Springs and later hearing a radio report, which I don't believe is accurate. . . . The 1994 PBS special *"American Experience"* features rare comments by FDR's attending physician, Dr. Howard Bruenn, by the time of his TV interview rather elderly, admitting to a few facts that conflict with the "official" Suckley story of Franklin's death, but rather confirming some of what Shoumatoff allegedly stated. . . . A "Bethesda source" posting, it was listed on AR15.com, stated his uncle was at the famous hospital and spoke to a man who was there, working at the time of FDR's death. It was learned - probably on 4/13/45 - that the beloved president died due to a self-inflicted wound, and when his aides "heard the shot and rushed into the room" they found him dead, *along with a suicide note,* which is a level of detail not seen in other postings. The note supposedly said that the president was upset about learning "the truth about the Communists," which sounds a little odd. But it was in response to another poster who said *he* heard the suicide theory from his stepfather, who in turn learned it from FDR insiders. . . . *"The Roosevelt Death"* book was copyrighted by G.L.K. Smith in 1947, or at least that year is stamped on the third page, and thus it is presumed this was the mysterious author's name. I read my copy of it online, thanks to resist.com. It is largely a hate-filled diatribe against "the dictator" FDR and distorts his policies, but at least the author/publisher was brave enough to discuss mostly unpopular opinions as far back as just two years from his death in a nation that largely revered the Roosevelt administration. . . . Curtis Dall quotes and facts come from his 1970 book, *"FDR, My*

Exploited Father-in-Law." Dall was largely excised by the family just as the Roosevelt administration began and his marriage ended. He had once been married to the president's daughter Anna, who helped arrange for Lucy Mercer to come back to her father towards the end of his life. I utilized *"Lies Your Teacher Taught You"* online for the book excerpts used herein. . . . Two amazing facts to consider: First, Eleanor Roosevelt surprised family and friends by refusing Secret Service protection in her widowhood, preferring to travel alone - with a handgun! Where she got the weapon seems to be a mystery, but the gun was almost always with her, to aid her defense if attacked (she never was, happily), and Mrs. Roosevelt even took some shooting and safety lessons in order to handle it properly. And second, when she confided to friend David Gurewitsch her private depression over her adult children's marital and professional woes in the years after Franklin's mysterious death, Eleanor let slip that "they brought her close to suicide," a 2014 book and TV special on the Roosevelts admitted. A *handgun*, and *suicidal thoughts*, for the troubled former First Lady? Now where did she get such notions? . . . A commentator of the contemporary FDR suicide allegations, Edmond Gunny, reminded all in an online forum: "Suicide is today often recognized as a moral resort to remove oneself from circumstances" that are considered unstoppable, upsetting, and overall "insurmountable." Could that have been the way FDR felt when he finished up his lifetime on earth? "This attitude was probably not as well received" in Roosevelt's day, added Gunny. It is better understood and sympathized with today, but in '45, for various reasons, such a tremendously embarrassing scandal *had* to be covered up, for everyone's good. . . . Added to the D.C. postmortem allegements as a near companion piece is a roundabout 2010 online allegation by a person who claimed he or she knew an Alabama man who explained he once met a traveling pharmaceutical salesman with a stunning story. The salesman assured the Alabaman that President Roosevelt did not suffer death by a stroke alone, "but it was caused by a bullet." The salesman said he knew this for a fact as he "had helped to prepare FDR for burial" (as part of the Spring Hill embalming group? Or at Bethesda? Or via a D.C. mortuary?) and saw the gunshot wound for himself. Obviously this is hearsay, and third-hand hearsay at that, but the claim matches perfectly. . . . Another web forum poster claimed in 2013 that he knew someone "whose cousin was FDR's personal assistant." This woman was in her forties at the time of the shocking death and was quickly "sent overseas." She evidently knew plenty of secrets. Supposedly never allowed to come back to America, the former aide found her letters censored and phone calls apparently monitored and blocked, plus "a permanent guard assigned to her." She allegedly managed to quickly let slip to a friend that she was undergoing all of this close scrutiny because she knew "FDR shot himself." . . . The poem "*The Lonesome Train*" was written by Millar Lampell, set to music by Earl Robinson in 1944. . . . FDR's friend and Under Secretary of State, Sumner Welles, had by 1945 resigned in scandal. He tried to carry on but also fell to excessive drink and depression. Years after Roosevelt's death, Welles too tried to kill himself, but survived the attempt. He carried on until his death in 1961, but this once again shows how far and how fast a well-known man in a position of great power in any age, in any nation, can swiftly fall from grace and health and impulsively desire his own end.

Bibliography

"*Above Top Secret: The Worldwide UFO Cover-up*," by Timothy Good, 1988, Morrow

"*Alien Agenda*," by Jim Marrs, 1997, Harper-Collins Publishers

"*Alien Contact: Top-Secret UFO Files Revealed*," by Timothy Good, 1991, Quill

"*Alien Mysteries, Conspiracies, and Cover-Ups*," by Kevin Randle, 2013, Visible Ink Press

"*Covert Retrieval: Urban Legend or Hidden History?*" 2013, by Linda L. Wallace

"*The Day After Roswell*," by Phillip Corso & William J. Birnes, 1997, Pocket Books

"*The Irregulars: Roald Dahl and the British Spy Ring in Wartime America*," by Jennet Conant, 2008, Simon & Schuster

"*It Didn't Start With Roswell: 50 Years of Amazing UFO Crashes, Close Encounters, and Cover-ups*," by Phillip Rife, 2001, Writers Club Press, an imprint of iUniverse.com, Inc.

"*The Lost Symbol*," by Dan Brown, 2009, Doubleday Publishing

"*Majic: Eyes Only*," by Ryan S. Wood, 2005, Wood Enterprises

"*Need To Know: UFOs, the Military, and Intelligence*," by Timothy Good, 2007, Pegasus Books

"*No Ordinary Time*," by Doris Kearns Goodwin, 1994, Simon & Schuster

"*Oppenheimer and the American Century*," by David C. Cassidy, 2005, Pi Press

"*Pearl Harbor: FDR Leads the Nation Into War*," by Steven M. Gillon, 2011, Basic Books

"*Rendezvous With Destiny*," by Michael Fullilove, 2013, Penguin Press

"*Rush Hudson Limbaugh*," by George Suggs, 2003, SEMO U. Press

"*Sage-ing While Age-Ing*," by Shirley MacLaine, 2007, Atria Books

"*Those Angry Days,*" by Lynne Olson, 2013, Random House

"*Truman*," by David McCullough, 1992, Simon & Schuster

"*UFO Crash/Retrievals: The Inner Sanctum, Status Report VI*," by Leonard Stringfield, 1991

"*UFOs: Myths, Conspiracies, and Realities*," by John Alexander, 2011, Thomas Dunn Books

"*What If*," by Shirley MacLaine, 2013, Atria Books

APX 1:

Scientist Vannevar Bush addresses President Truman about FDR's desire to learn more about "non-terrestrial science" in a July 1947 memo that mentions "material fabrication" of the spaceships recovered.

RECEIVED

USED TOP SECRET
VANNEVAR BUSH FILE
OPERATION MAJESTIC TWELVE **4102I**
LETTER OF TRANSMITTAL

Office of Scientific Research and Development
1530 P Street, NW.
Washington 25. D.C.

July 5. 1947

Dear Mr. President:

In a April 1944 letter. President Roosevelt requested my recommendations on the following points:

(1) What can be done, consistent with military security, and with the prior approval of the military authorities, to make known to the P-3 program, as soon as possible the contributions which have been made during our war effort to scientific knowledge?

(2) With particular reference to the war of science against foreign technology, what can be done now to organize a program for continuing in the future the work which has been done?

(3) What can the Government do now and in the future to aid research activities by private organizations?

(4) Can an effective program be proposed for discovering and developing scientific talent in America so that the continuing future of scientific research in this country may be assured on a level comparable to what has been done during the war?

It is clear from President Roosevelt's letter that in speaking of science he had in mind the non-terrestrial sciences, including biology and physics, and I have so interpreted his questions. Progress in other fields, such as material fabrication, is likewise important; but the program for non-terrestrial science in my report warrants immediate attention.

In seeking answers to President Roosevelt's questions I have had the assistance of distinguished committees specially qualified to advise in respect to these subjects. The committees have given these matters the serious attention in light of recent developments; indeed, they

TOP SECRET/MAJIC EYES ONLY
OPERATION MAJESTIC 12

ORIGINAL

APX 2

In 1999, retired CIA and Army Counter Intelligence officer Thomas Cantwheel confessed in a typed letter that one of the government's top secrets was a recovered spaceship from Missouri in 1941, and that it was replicated in a secret experimental aircraft.

4-15

THE "LOANED" AIRCRAFT WAS ACQUIRED IN 1945 FROM THE AIR TECHNICAL
SERVICES COMMAND (ATSC, NOW AEDC, AIR FORCE SYSTEMS COMMAND).
THE "S" AIRCRAFT WAS WAS DESIGNED FROM A AERODYNE RECOVERED IN
1941 THAT CRASHED IN SOUTHWESTERN MISSOURI AND ONE CAPTURED IN
1942 IN LOUSIANA. RE-CONSTRUCTION COMMENCED IN 1945 WITH THE
ASSISTANCE OF GERMAN SCIENTISTS AT WRIGHT FIELD. PROPULSION
PROGRAMS TRIED TO DUPLICATE THE ATOMIC POWER PLANT FOUND ON THE
AERODYNE CAPTURED IN LOUSIANA AND INTEGRATED THE MAGNETIC DRIVE
SYSTEM DEVELOPED BY TESLA. THE AAF HOPED TO INTRODUCE THIS CRAFT
IN WAR, BUT RESOURCES AND MONEY WAS NOT AVAILABLE. FUNDING WAS
NOT AVAILABLE TO R&D UNTIL 1946, WHEN LEMAY TOOK OVER THE PROJECT.
THE S CRAFT COULD TAKE OFF VERTCALLY AND REACH ALTITUDES AS HIGH
AS 90,000 FEET AT SUPERSONIC SPEEDS. THE CONTROLS WERE ELECTRONIC
FLY BY WIRE AND THE PILOT LOOKED AT PROJECTION OF SYMB OLS ON A
TRANSPARENT SCREEN. THE CRAFT WAS SO COMPLICATED THAT TEST PILOTS
HAD GREAT DIFFICULTY IN HIGH PERFORMANCE AT VERY HIGH ALTITUDES.
SEVERAL TEST PILOTS WERE KILLED AS A RESULT OF DECOMPRESSION AND
EJECTION CAPSULE OR ESCAPE CYLINDERS WERE NOT DESIGNED FOR HIGH
ALTITUDE EJECTIONS. AS A RESULT, ONE "S" CRAFT WAS LOST. THE
MATERIALS USED IN THE CONSTRUCTION BY HUGHES AIRCRAFT COMPANY
FAILED TO PROTECT THE TEST PILOTS IN MAXIMUM THROTTLE SETTINGS
AND EXPOSED THEM TO HIGH DOSE OF RADIATION WHICH RESULTED IN
SERIOUS ILLNESS AND DEATH. TEST FLIGHTS CONTINUED OVER THE WSPG
IN EARLY 1947 AT KIRTLAND AAF, AND AT ALAMOGORDO AAF TULAROSA
RANGE WITH BETTER RESULTS TO FLIGHT PERFORMANCE, BUT EXPOSURE
TO THE RADIATION FROM THE ATOMIC ENGINE CONTINUED. IN 1947,
THE "S" CRAFT WAS MODIFIED TO CARRY ATOMIC WEAPONS OVER LONG
DISTANCES. BUT HIGH ALTITUDE FLIGHT SIMULATIONS AND PILOT
SURVIVABILITY WAS STILL A PROBLEM. BELL AIRCRAFT WAS GIVEN THE
CONTRACT FOR A ROCKET-POWERED AIRCRAFT TEST BED TO RECORD FLIGHT
DATA AT SUPERSONIC SPEEDS AS A PROBLEM SOLVING LABORATORY. DATA
COLLECTED FROM X-1 AIRCRAFT WAS UTILIZED FOR DESIGN MODIFICATION
OF "S" CRAFT AERODYNAMIC TESTS AT LANGELY. RADIATION PROBLEMS
PROMPTED A CHANGE IN THINKING IN THE TEST PILOT PROGRAM PROMPTED
A NEW PILOT "MODEL" EXPERIMENT USING ALTERNATE CHOICE OF SUBJECTS
WAS APPROVED. THIS R&D PROGRAM WAS KEPT SECRET FROM OTHER COMMANDS
AND SERVICES, AND WAS GIVEN SPECIAL STATUS EQUAL TO THE SECURITY
OF THE ATOMIC BOMB PROGRAM. AT LEAST A DOZEN "S" CRAFT WERE BUILT
AND TEST FLOWN. THREE WERE LOST DUE TO MECHANICAL FAILURE AND
PILOT ERROR. TWO MORE "S" CRAFT WERE LOST ALONG WITH FIVE FATALIZIES,
THAT CAUSED THE AAF TO CANCELL THE PROJECT INDEFINITELY. ALL SIMILAR
"WING" AIRCRAFT PROJECTED FOR SERVICE IN AAF WERE LIKEWISE CANCELLED
IN 1949 AT THE RECOMMENDATION OF GENERAL LEMAY. FUTURE ATOMIC-POWERED
AIRCRAFT WERE "PILOTLESS" AND WOULD BE CONTROLLED REMOTELY. THE AF
WOULD NOT RESUME TESTING OF "LOANED" AERODYNES UNTIL PILOT SAFETY
ISSUES WERE SATISFIED AND A MORE SECURE TEST RANGE COULD BE USED
FOR AEROSPACE R&D PROGRAMS. AREA 51 IS BUT ONE SUCH RANGE AMONG
THE SECRET TEST SITES NOW IN OPERATION. A SPECIAL RECOVERY AND SECURITY
UNIT CALLED S.T.U.D.S. WAS CREATED IN THE LATE 1980's TO FACILITATE
THE SERVICING OF NEWER "S" CRAFT FLIGHT TEST OPERATIONS. EFFORTS TO
CONCEAL THE TRUE NATURE OF FLIGHT OPERATIONS WERE SUCCESSFULL IN
THAT THE AF DEVISED A COVER INTELLIGENCE PROJECT CALLED BLUE BOOK.
PROJECT BLUE BOOK, AS A COVER PROJECT, WAS CONTROLLED BY THE CIA
TO PROTECT AF TEST FLIGHT OPERATIONS FROM SPECULATION BY THE PUBLIC,
AND CONVINCE THE SOVIETS THAT USAF HAD NO AIRCRAFT CAPABLE OF FLIGHT
CHARACTERISTICS AND MANSOVERS AS OBSERVED AND REPORTED TO BLUE BOOK,
AND THE USAF UFO PROGRAM. IN 1958, PROJECT UFO AND MOON DUST WERE
ACTIVATED WHEN USA INTERPLANETARY PHENOMENE UNIT OPERATIONS CEASED
AND CIC RESPONSIBILITY FOR UFO SECURITY WAS TRANSFERED TO USAFOSI.

APX 3

President Roosevelt's February 1944 Oval Office memo to his "Non-Terrestrial Science and Technology Committee," about how to handle the MO41 recoveries - and weaponize them.

DOUBLE TOP SECRET   February 22, 1944.

THE WHITE HOUSE

WASHINGTON   February 21, 1944

MEMORANDUM FOR

THE SPECIAL COMMITTEE ON NON-TERRESTRIAL SCIENCE
AND TECHNOLOGY

agree with the O.S.D proposal of the recommendation put forward by Dr. Bush and Professor Einstein that a separate program be initiated at the earliest possible time. I also agree that application of non-terrestrial know how in atomic energy must be used in perfecting super weapons of war to affect the complete defeat of Germany and Japan. In view of the cost already incurred in the atomic bomb program, it would, at this time, be difficult to approve without further support of the Treasury Department and the military. I therefore have decided to forego such a enterprise. From the point of view of the informed members of the United States, our principle object is not to engage in exploratory research of this kind but to win the war as soon as possible.

Various points have been raised about the difficulties such an endeavor would pose to the already burdened research for advanced weapons programs and support groups in our war effort and I agree that now is not the time. It is my personal judgement that, when the war is won, and peace is once again restored, there will come a time when surplus funds may be available to pursue a program devoted to understanding non-terrestrial science and its technology which is still greatly undiscovered. I have had private discussions with Dr. Bush on this subject and the advice of several eminent scientists who believe the United States should take every advantage of such wonders that have come to us. I have heard the arguments of General Marshall and other members of the military that the United States must assume its destiny in this matter for the sake of the Nation's security in the post-war world and I have given assurances that such will be the case.

I appreciate the effort and time spent in producing valuable insights into the proposal to find ways of advancing our technology and national progress and in coming to grips with the reality that our planet is not the only one harboring intelligent life in the universe. I also commend the committee for the organization and planning that is evident in Dr. Bush's proposal and the delicate way in which it was presented. I trust the committee will appreciate the situation on which this office must render its decision.

[signature]

DOUBLE TOP SECRET

APX 4

President Roosevelt's February 1942 Oval Office memo to General George Marshall regarding better access to the MO41 alien recoveries.

TOP SECRET

Top Secret

gm 25
February 27, 1942

THE WHITE HOUSE
WASHINGTON

February 27, 1942

MEMORANDUM FOR

CHIEF OF STAFF OF THE ARMY

     I have considered the disposition of the material in possession of the Army that may be of great significance toward the development of a super weapon of war. I disagree with the argument that such information should be shared with our ally the Soviet Union. Consultation with Dr. Bush and other scientists on the issue of finding practical uses for the atomic secrets learned from study of celestial devices precludes any further discussion and I therefor authorize Dr. Bush to proceed with the project without further delay. This information is vital to the nation's superiority and must remain within the confines of state secrets. Any further discussion on the matter will be restricted to General Donovan, Dr. Bush, the Secretary of War and yourself. The challenge our nation faces is daunting and perilous in this undertaking and I have committed the resources of the government towards that end. You have my assurance that when circumstances are favorable and we are victorious, the Army will have the fruits of research in exploring further applications of this new wonder.

    You may speak to me about this if the above is not wholly clear.

F. D. R.

TOP SECRET

This is a retyped version of FDR's memo from DE's DS files

APX 5

General Nathan Twining's September 1947 "White Hot" report reference to "the recovery case of 1941," in other words the Cape Girardeau spaceship crash.

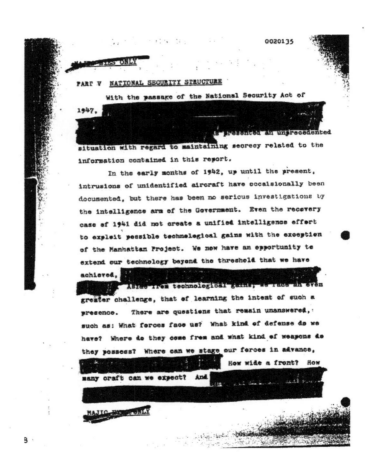

0020135

~~MAJIC EYES ONLY~~

PART V NATIONAL SECURITY STRUCTURE

With the passage of the National Security Act of 1947,

is presented an unprecedented situation with regard to maintaining secrecy related to the information contained in this report.

In the early months of 1942, up until the present, intrusions of unidentified aircraft have occaisionally been documented, but there has been no serious investigations by the intelligence arm of the Government. Even the recovery case of 1941 did not create a unified intelligence effort to exploit possible technological gains with the exception of the Manhattan Project. We now have an opportunity to extend our technology beyond the threshold that we have achieved,

Aside from technological gains, we face an even greater challenge, that of learning the intent of such a presence. There are questions that remain unanswered, such as: What forces face us? What kind of defense do we have? Where do they come from and what kind of weapons do they possess? Where can we stage our forces in advance, How wide a front? How many craft can we expect? And

~~MAJIC EYES ONLY~~

B

About the Author

Paul Blake Smith was born and raised in MO41's hometown of Cape Girardeau, Missouri. He is a product of that city's public school system, and then attended Southeast Missouri State University in Cape, where he was a four-year Mass Communications Major with an English Minor. The grandson of a Cape Girardeau attorney of fifty-plus years - and also a U.S. Commissioner, City Attorney, and community leader – Paul is the son of a local paralegal and an educator who worked in the suspected crash area within neighboring Scott County. A fan of American history and popular culture, Paul now lives and works in a city in the western part of the Show-Me State, writing original screenplays and other books on largely historical, nonfiction subjects – including a follow-up to this original fact-based publication. **MO41, The Bombshell Before Roswell** is his first published book.

Printed in Great Britain
by Amazon

42785006R00190